ADIRONDAC
P9-APJ-765
BAY ROAD

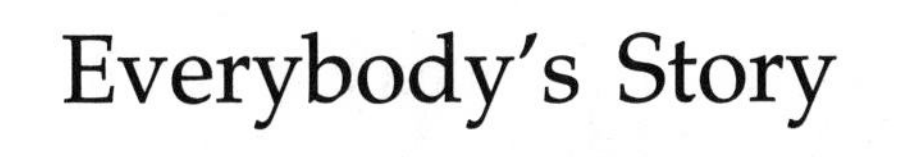
Everybody's Story

SUNY series in Philosophy and Biology
David Edward Shaner, Editor

Everybody's Story

WISING UP TO THE EPIC OF EVOLUTION

Loyal Rue

With a Foreword by
Edward O. Wilson

STATE UNIVERSITY OF NEW YORK PRESS

Cover art adapted from *Those People are Like Animals* by S. Mindrum–Logan.
Interior Illustrations by Richard Merritt.

Published by

State University of New York Press, Albany

Printed in the United States of America

For information, address
State University of New York Press
University Plaza, Albany, NY 12246

Production by Michael Haggett
Marketing by Patrick Durocher

Library of Congress Cataloging-in-Publication Data

Rue, Loyal D.
Everybody's story : wising up to the epic of evolution / Loyal Rue : with a foreword by Edward O. Wilson.
p. cm. — (SUNY series in philosophy and biology)
Includes bibliographical references and index.
ISBN 0-7914-4391-4 (alk. paper). — ISBN 0-7914-4392-2 (pbk. : alk. paper)
1. Evolution—Religious aspects. 2. Ethics, Evolutionary.
I. Title. II. Series.
BL263.R812 2000
146′.7—dc21 99-13703
CIP

10 9 8 7 6 5 4 3 2 1

To Marilyn, Carl, Anna, and Elena,
without whom I would have less care for the future

Contents

Foreword by Edward O. Wilson ix

Preface xi

Introduction 1

Part I: How Things Are 45

1. The Organization of Matter 53
2. The Organization of Life 65
3. The Organization of Consciousness 81

Part II: Which Things Matter 97

4. What Matters Ultimately? 99
5. What Matters Proximately? 109

Epilogue 129

Bibliographical Notes 139

Index 143

Foreword

Homo sapiens can justly be called the mythopoeic species. Human beings must have an epic, a sublime account of how the world was created and how humanity came to be part of it. The brain's architecture automatically makes up stories; and the mind it creates is a theater of competing scenarios. The brain is not confined, animal-like, to instant sensory impressions followed by rough associations of these impressions with past reward and punishment. Instead, it searches continuously backward across time to re-create past events, real and imaginary, and forward to invent future scenarios. Stories that are pleasing to reason and emotion outcompete others less so. Replacing them, they serve thereafter as maps of future action. During this process the self, the central protagonist of the scenarios, is perceived within the present-moment scenario as having reached a decision.

The primal instinct of narrative, of continuous scenario invention, is what makes the human brain superior in performance. In dreams we construct stories of unconstrained fantasy. In gossip we evaluate others with tales of their exploits and foibles. And in religious myths we repeat the epics that ennoble our lives, our tribe, and our species.

Religious epics satisfy another primal need. They confirm that we are part of something greater than ourselves. They say,

Death may claim your precious self, and those you most love, but it will not claim the tribe or sully the benefits that empower the tribe.

To have credibility, the religious epic must be thought superior to the stories of competing tribes. Even to think otherwise is, in the eyes of fundamentalists, heretical, blasphemous, and traitorous. Superiority of one's religious epic is a sacred imperative.

To justify religious epics, the two connate properties of human nature, the narrative and spiritual drives, have always served to divide humanity. They create a terrible dilemma: How are we to satisfy them, even enrich them, without the continuance of falsehoods that promote divisiveness and conflict? Is there a way to evolve a great epic that is at once universal, spiritually satisfying, and, above all, truthful?

The quest for such an epic is the subject of *Everybody's Story*. Loyal Rue's argument is as bold as it is brief: The way to achieve an epic that unites humanity spiritually, instead of cleaving it, is to compose it from the best empirical knowledge that science and history can provide of the real human story. Spirituality is beneficent to the extent that it is based on verifiable truth.

I find his argument persuasive.

Edward O. Wilson
Pellegrino University Professor
Harvard University

Preface

> What new story can explain the relationship between human civilization and the earth—and how we have come to a moment of such crisis? One part of the answer is clear: our new story must describe and foster the basis for a natural and healthy relationship between human beings and the earth.
>
> —Albert Gore, Jr., *Earth in the Balance*

It can be no mere coincidence that cultures everywhere employ stories when they turn to the important business of guiding youthful minds toward a full life. The universal practice of storytelling says that at some level there is a recognition that human lives are themselves narrative realities, each one joining the stream of an ever-enlarging story. From the very beginning we humans nurture and orient our children with stories, drawing them into a reality of enduring promise. I am certainly not aware of every life-orienting story to emerge in the history of human experience. Of those I do know, however, there are none failing to stress the point that life itself is a gift of incomparable worth, always to be appreciated and never to be taken for granted.

My own experience has been no exception. As a child I attended Sunday school faithfully, there to hear stories from the Bible that invariably left me with the sense that I was blessed, that the life I had was not of my own making but rather a gift of transcendent grace. After hearing the Bible stories we children would respond together in singing the following verse:

> I love to tell the story
> Of unseen things above,
> Of Jesus and his glory,
> Of Jesus and his love.
> I love to tell the story,
> Because I know 'tis true;
> It satisfies my longings
> As nothing else can do.

But that was then. Things are quite different now, even though some of the formal elements have remained the same. I still possess a deep sense of gratitude for the life I have, and this sense is still derived from the features of a story. And indeed I still love to hear and to tell the story. But the story itself has changed. The reality of enduring promise into which I am now drawn has no longer to do with "unseen things above." It is instead the reality described in broad terms by the scientific narrative of our cosmological journey, the epic of evolution. The epic of evolution is the sprawling interdisciplinary narrative of evolutionary events that brought our universe from its ultimate origin to its present state of astonishing diversity and organization. In the course of these epic events matter was distilled out of radiant energy, segregated into galaxies, collapsed into stars, fused into atoms, swirled into planets, spliced into molecules, captured into cells, mutated into species, compromised into ecosystems, provoked into thought, and cajoled into cultures. All of this (and much more) is what matter has done as systems upon systems of organization have emerged over fifteen billion years of creative natural history.

This epic of evolution is the biggest of all pictures, the narrative context for all our thinking about who we are, where we have come from, and how we should live. It is the ultimate account of how things are, and is therefore the essential foundation for discourse about which things matter. I am moved to tell this new story partly because I have found it deeply satisfying as a resource for self-understanding. To know one's place in the cosmos is to know something of immense spiritual value. But more than this, I tell it from the conviction that our knowing the epic of evolution bears on the evolution of the epic. There is an element of Socratic doctrine here: our knowing affects our doing.

But knowing is never enough. To act on what we know requires that we also feel. My Sunday school teachers did not merely inform us about Bible stories—they told the stories to educate our emotional lives as well, to engage our affections that we might serve the enduring promise. This book shares their view that we will not serve what we do not love. And we cannot love what we do not know. The story told here is therefore an invitation to know and to love the fifteen-billion-year-old process that has blessed us with the lives we have. To know this process is to love it.

And to love it is to serve it in whatever measure we are able. We human beings have much to answer for in the way we have assaulted the integrity of the earth's natural systems. There will be much for loving servants to accomplish in the critical generations ahead. It seems clear that the children born at the turn of the millennium are already destined to endure a lifetime of great suffering and/or a lifetime of deep systematic change. It is for their sake that this book is written. They will need the resources to create a new world of solidarity and cooperation. And first among these resources is the unifying power of a common story, everybody's story.

Here, then, is what I intend for this book: to participate in the important work of constructing a new wisdom tradition that couples an evolutionary cosmology to an ecocentric morality. The specific aim of this book will be to tell the story of the universe in a manner that might inspire grateful service to the enduring promise of life on the planet. In the broadest sense, this book is a modest attempt at helping us to get our heads and hearts working together so that we can feel a bit easier about future generations.

The structure of the book is simple and straightforward. First comes a fairly long introduction explaining what story traditions are, how they work, and why we should start a new one. Here I argue that wisdom traditions, that is, story traditions, arise from an integration of fundamental facts and ultimate values. These traditions have the power to organize our individual and collective lives by educating our thinking and feeling. Wisdom traditions, however, are not eternal—they emerge out of crisis, they flourish for an age, and they begin to falter under challenges to their intellectual and moral authority. As they decline, so does the solidarity and cooperation they once fostered. If circumstances reach a point of crisis, a new story will arise.

Then comes the story itself, the epic of evolution. Part I narrates "How Things Are." Chapter 1 gives an account of the evolution of matter, chapter 2 traces the evolution of life, and chapter 3 considers the evolution of consciousness. Each of these chapters concludes with a reflective interlude. Part II then presents a view of "Which Things Matter." Chapter 4 argues that the continuation and fulfillment of life is an ultimate value. Chapter 5 discusses the proximal values of personal wholeness, social coherence, and the integrity of the earth.

The epilogue considers the prospects that a new wisdom tradition might arise to offer adequate resources for the achievement of global solidarity and cooperation. One of the important issues raised by the emergence of a new wisdom tradition is the extent to which it will impact on existing religious traditions. Will the established religions of the world adjust to the epic of evolution, thereby transforming themselves in the process? Or will they gradually be displaced by emerging forms of religious life that take their inspiration more directly from the epic of evolution? No doubt it is too soon to tell. But it is clearly not too soon to tell that this dilemma will continue to be a source of tension within the spiritual lives of individuals for generations to come.

Introduction

The Challenge of Wising Up

Oliver Sacks tells a story about a patient of his who had been blind since birth and first received the gift of eyesight in middle age, following a delicate operation. The patient's wife phoned Sacks immediately after the operation, eager to share the good news that her husband could see at last. Sacks asked what her husband saw. The woman replied that although he could see perfectly well he had no idea *what* he was seeing. The clinical explanation for this unusual experience is that the brain of the newly sighted man was not conditioned neurologically to interpret the stark novelty of visual sense data. Information was flowing in but he had no way to give it any meaning.

The condition of the newly sighted man speaks for the perplexities of historical existence. We human beings find ourselves shaped by the past to make sense of the circumstances of the past. But when it happens that circumstances change radically, our ways of understanding and our modes of response are rendered inadequate. Such is the present condition of our species. Our attempts to comprehend the problems now experienced on a global scale are as tentative as the attempts of a newly sighted man to interpret unfamiliar sense data. Our brains are unprepared for the complexity.

The problems at hand are immense, complex, and unprecedented. Nevertheless, we are able to see that the fundamental response they demand of us is both straightforward and familiar. That is, we are called to enlarge our sense of human solidarity and to create bold new means of cooperation. If we are to hope for a viable future on this delicate planet then we must in some measure learn to think and to act as one species.

This book takes seriously the proposition that we are now, as a species, poised for a transformation of consciousness, a process of the utmost difficulty having consequences of the utmost significance. Yet the challenge should not be overdrawn. We are called to *transform* consciousness, not to invent it. Thus our first step should be to take our bearings, using whatever timeless insights we might bring from the past. Certainly the most important human insight of all time is expressed in the imperative to live in harmony with reality, a principle fundamental to the life process itself. To live in harmony with reality is to have a fighting chance. But if we live at odds with reality then the odds are that we shall be prematurely swept or worn into oblivion. This timeless insight is not a terribly complicated idea, yet it is no less profound or significant for all its simplicity. In fact, I suggest it is this very insight that we humans bear in mind when we use the term "wisdom." Wisdom *just is* a way of thinking that puts the odds in our favor. Wisdom is the intellectual and moral wherewithal to live in harmony with reality. Nothing, surely, could be more germane to human interests than the quest for wisdom.

It is commonly said that this quest begins in a state of wonder, the surprised *dis-ease* that overtakes one whose assumptions and expectations have been undone by events. Imagine yourself drawing on an inside straight in a game of poker and the card you turn up happens to be a blue eleven of stars! Immediately you are plunged into a state of wonderment. This cannot be, you say. But *is* it? You take a second look. Sure enough! But what can this mean? Something is amiss and urgently needs sorting out. Wonderment, like the disturbance of new sight, invites a reorganization of consciousness.

Socrates understood matters of wonder and wisdom. He was reputed to be the wisest man in Athens, but to his mind this meant merely that he alone was aware of the extent to which he was unwise. For Socrates the first mark of wisdom is an honest recognition of human folly. It is in this self-critical spirit of Socratic wonder that wisdom for a new millennium must be sought.

Folly and Distress

The extent of our folly is already a familiar theme: human beings are living on the brink of self-induced disaster. We have created a world in which the natural and social systems giving us life are being stressed to the point of collapse. We know the problems by heart: global warming, ozone depletion, the extinction of species, soil erosion, toxic waste, air and water pollution, mineral and fossil fuel depletion, poverty, crime, injustice, terrorism, exploitation—the list goes on and on. Gerald Barney has popularized the term "global problematique" to refer to the cumulative challenge of these interrelated problems:

> As we humans have begun to think globally, it has become clear that we do not have a poverty problem, or a hunger problem, or a habitat problem, or an energy problem, or a trade problem, or a population problem, or an atmosphere problem, or a waste problem or a resource problem. On a planetary scale, these problems are all interconnected. What we really have is a poverty-hunger-habitat-energy-trade-population-atmosphere-waste-resource problem. (Barney, p. 25)

One species, one world, one problem. These are the realities challenging us to be newly wise. The general features of our planetary megaproblem are just now coming into focus. The global problematique is global, systematic, immediate, and chronic. The problem is *global* in the sense that its causes and consequences are to be shared by virtually everyone on the planet. The finger of blame points everywhere and there is nowhere to hide from the penalties. It is a *systematic* problem in the sense that it has many interconnected parts. Disruptions to a natural or social system in one part of the world can be expected to have multiple unforeseen consequences for other systems, sometimes remote in time and space. Whether we like it or not, the life we have is dependent on a global swarm of interlocking systems. Whatever happens here and now will matter, in some sense, everywhere and forever. The problem is *immediate* in the sense that it is in our faces right now—the victims are already with us. Scientists tell us that the earth's climate systems have already been irreversibly altered, and that a mass extinction of plants and animals is well under way. And the problem is *chronic* in the sense that it will dominate the human agenda for the indefinite future. No person alive will remain unaffected, but it is clear that the most severe consequences await those who are not yet born. The megaproblem will persist because the driving forces behind it—beliefs, policies, traditions, technologies, habits, hopes, values, ignorance—are all lined up to resist fundamental change.

A composite scenario may help to illustrate the general features of our global problematique. Picture a third world nation, call it *Poorlandia*. Poorlandia has been dominated for centuries by a religious tradition that places women in a subordinate position, granting them very few educational or economic opportunities. As a result, the social esteem of Poorlandian women is linked to their performance as childbearers. The economic status of Poorlandian families may be enhanced by having many children. Children can be put to work bringing resources into the family. And as parents grow old they will come to depend on their children as the sole form of social security. All of which means it is economically foolish not to have a large family. So the population grows. Industrial nations begin to take interest. Some industrialists view Poorlandia as a fresh market for first-world products, others see a source of cheap resources. The industrialists are driven to seek out new markets and cheap resources by the logic of their own economic system. If you live in a third world nation it may be perfectly rational to have lots of kids, and if you live in an industrial nation it will make perfect sense to get out there and develop new markets and exploit new resources.

Poorlandia is an impoverished and unsophisticated country, vulnerable to manipulation by first-world developers. So the nation accepts a massive loan along with a complicated development scheme. Poorlandia's political leaders are corrupt, so most of the loan money goes into the pockets of opportunists rather than into long-term economic development. But the loans must be repaid eventually, so the country increases incentives for local developers to come in and cut down the forests. Timber is sold to industrial consumers and the land is cleared for cattle ranching to supply beef to fast-food chains in industrial nations (where business is booming because first-world families can't be bothered to shop for groceries and cook for themselves). The impact on natural systems is severe. Habitat for countless species is destroyed by clear-cutting forests. Clearance fires pump carbon dioxide into the atmosphere, adding to the process of global warming. The fragile topsoil goes barren after two or three years, increasing demand for more pastureland. In order to keep up with inflation rates and loan payments Poorlandia sells dumping rights for toxic waste generated by the first world. Indigenous populations are forced out of subsistence to go live in teeming urban ghettos—there to remain jobless and uneducated, yet able to produce (and unable to prevent) lots of children. The children, to support themselves and their families, turn to crime and prostitution.

These are a few of the dynamics of the global problematique. The cumulative impact of such interlocking forces has been to stress natural and social systems on a global scale. Humans are presently coopting the resources of the planet at an alarming rate, recklessly destroying ecosystems and natural habitat well beyond the ability of nature to repair and replenish. And social systems are sent topsy-turvy in the process. In the contemporary world, once-reasonable behavior has become grossly irrational, once-virtuous solutions now produce vicious problems, what was once wisdom has become radical folly.

If we ask how our world came to be in such a state the answer is not hard to find. The various features of the global problematique can be traced directly to the combined effects of excessive reproduction and material consumption among human beings. And if we ask how these patterns of excess came to pass we can point to the triumph of human invention over the elements of nature. For the vast majority of our species' history the contest between humans and nature was at equilibrium. But in the course of events we have ingeniously thwarted the major factors (e.g., disease, predation, food shortage) that once kept our population relatively stable. This, however, is merely the population part of the formula.

We owe the bulk of our excessive consumption (and pollution) to a range of pursuits having no direct bearing on the goal of survival. It is clearly the case that the vital needs of the earth's six billion people add up to no more than a mere fraction of the consumption and pollution presently stressing the planet's carrying capacity. Mere survival, it seems, is not sufficient for a species clever enough to clobber the odds against it. So we take our talents and tastes, endowed by natural history for their survival value, and use them to invent an endless menu of spare-time pursuits—and these we encourage under the ideals of social progress and personal success. Of course, many social and personal goals are relevant to the challenge of survival. If we want the security of continued access to vital resources we shall take both individual and collective measures to stockpile goods and to protect them. But far beyond what is justified in the name of security, contemporary ideals of social progress and personal success have become instruments for encouraging the excesses of an overgrown culture of consumption.

The prevailing standard of social progress is not difficult to comprehend. It is simply a measure of the degree to which a group is able to create the conditions for personal success. Progressive nations are, therefore, aggressive ones—those managing to assemble commercial, military, political, and technological elements into formulas for economic growth. Economic growth means that conditions at home will be generally favorable for individuals to advance their goals for personal success. The standard for personal success is not complicated either. It measures the degree to which individuals have the means to get whatever they want. That is, whatever their expansible needs for self-esteem dictate—which can become inordinate in a media-dominated culture that glorifies self-indulgence and consumption.

I stress the point that these are *prevailing* standards of social progress and personal success. There is no reason in principle that collective and individual ideals cannot be formulated in ways that do not encourage economic growth and material consumption. Nevertheless, for the past two centuries it has become increasingly conventional to judge national and personal worth by unsustainable economic standards.

A major factor in the global problematique is that industrial economies have developed in ways that make it virtually impossible to distinguish effectively between essential consumption (i.e., relative to the basic needs of life) and extraneous consumption (i.e., relative to nonessential, trivial pursuits). This is a significant problem because it is excess in the latter that is primarily responsible for environmental distress. The distinction is promising because it suggests the possibility of addressing a majority of our environmental problems by simply discouraging nonessential industries. A moratorium on trivialities, so to speak. A nice solution, except that it ignores the systematic manner in which the world's economies have coupled the essential to the extraneous. That is, for a large share of the world's population the available means for acquiring essential goods are linked to the production of trivial goods. For example, even though there may be nothing more frivolous than a video arcade, if I am employed by one I will view it as a vital means to acquire food and shelter.

The folly of our species is undeniably manifest to anyone willing to open their eyes. We have blindly pursued unrealistic goals of progress and success, pursuits that have led to an explosion of human population and a frenzy of material consumption. The consequences of our folly, however, are painfully real—our viability as a species has been brought to point by the distress we have imposed on life-giving natural and social systems.

The Range of Options

Finding a way to think about solutions to the global problematique is not all that difficult. That is, it turns out there are not a whole lot of options. Five to be exact. The first option is to do nothing and let nature take its course. This is the *sit-back option.* On this option we just try to ignore the problems while charging forward in pursuit of more progress and success. The odds of our following this option look pretty strong given the reluctance of human beings to welcome radical change. So we might simply stay the course of folly until nature imposes a massive correction, perhaps in the form of widespread starvation or disease. Each of these devastators becomes more likely as human population continues to rise. At sit-back rates the population is expected to soar to eleven billion by the year 2050. This figure would bring the demands for nourishment well beyond the earth's potential for supply. At sit-back rates more than 25 percent of the earth's plant and animal species will be extinct by 2150. Under such conditions life at the top of the food chain would become extremely precarious. Nobody can predict the sit-back rates for disease, but we do know that as population density increases so does the likelihood of unleashing new vectors of virulent microscopic killers.

If we are not content merely to sit back there remains the *spread-out option.* This is an ancient strategy, practiced routinely by our hunting-gathering ancestors for hundreds of thousands of years. The idea is simple: when population density increases to the point of stressing natural and social systems then a portion of the population relieves the stress by packing off to new territory. The modern term for this strategy is *colonialism.* The colonialist option might be an appealing solution to the global problematique if it were not for the fact that it has been a principal cause of it. Poverty and overpopulation in the third world together with exploitation and overconsumption by the industrialized world are the legacies of colonialist policy. The spread-out option has resulted in scores of assaults on natural and social systems. The trouble is that once limits have been reached, spreading out no longer brings relief from stress, it just means more stress. Of course the subject of limits is not a source of worry for those infatuated with the prospects for colonizing outer space. Thus we are subjected to proposals for building metropolitan space stations and for seeding life forms on some of the thousands of asteroids in our solar system—the ultimate triumph of colonialist mentality. Despite the fact that some highly respected and competent individuals have promoted the colonization of space, there remains no good reason to consider these projects anything more than idle fantasies feeding off a fallacious analogy between the new world of the fifteenth century and the open sky of the twenty-first.

A more realistic alternative (though hardly more appealing) is the *kill-off option*. If the supply of space and resources cannot be increased then perhaps the demand for them might be decreased—and this could be achieved rather quickly by culling out a substantial fraction of the human population. The conventional means for this option are warfare and infanticide, both of which have been practiced with regularity throughout the history of the species. Despite these precedents it will be difficult for many to view the kill-off option as anything but a failure of social and political imagination. Organized murder is not a solution to anything, it is a desperate reaction to intractable problems. And given the technology of modern warfare the kill-off option, like the spread-out option, promises only to make the problems infinitely worse.

The kill-off option offers a response to the global problematique by reducing the number of individuals making demands on the environment. An alternative to this strategy would be to reduce the level of each person's demands. Thus we get the *hunker-down option*. This option exhorts individuals to consume less in order to make way for an increasing population. Ask yourself how you might live if your income were to be cut in half. That's what it would be like to hunker down. Hunkering down is what people typically do during wartime, when goods become scarce—they cut out the luxuries, water down the soup, make do. The hunker-down option intervenes on the consumption side of the problem. It is a noble and necessary part of any response to the global problematique, but as long as human population continues to rise this option can do little more than buy time. Moreover, it is not clear that the hunker-down option can be sustained. It is too ascetic. Most individuals are prepared to make sacrifices while under temporary siege, but as their endurance fades they begin to look with favor to the options for spreading out and killing off. Getting people to hunker down as a permanent way of life would require deep systematic changes.

This book favors a fifth alternative, the *wise-up option*. The wise-up option opposes the denial of the sit-back option, the folly of the spread-out option, the desperation of the kill-off option, and the austerity of the hunker-down option. It takes seriously the limits of natural systems and seeks the social and psychological means by which our species may live sustainably within them. The wise-up option seeks a new way to live in harmony with reality.

In various measures and combinations all five of the above options have been taken by human beings during critical periods of stress in the past. Of these strategies the wise-up option has been the least common, perhaps because it is the least obvious and the most radical. Yet it is the one option that is exclusive to our species. Other species can sit back, spread out, kill off, and hunker down. But humans alone have the ability to wise up. To see more clearly what is meant by wising up it will be useful to focus briefly on the foundations and development of human social existence.

On Human Nature and History

Somewhere between three and four thousand distinct cultural traditions have appeared in the history of our species. The vast diversity of these traditions can, however, be reduced to just a few basic types of social organization, emerging in historical succession. The point I wish to stress is that our wising up from one form of social organization to another has always been achieved by manipulating those aspects of human nature that are most conducive to the expansion of solidarity and cooperation, our emotional lives in particular.

Nature has endowed all social species with rough measures for deciding whom to help or to spurn. In higher primates *kin selection* and *reciprocal altruism* are the principal mechanisms for determining appropriate behaviors toward other members of our own species. These mechanisms amount to innate rules guiding us to act in certain ways under certain circumstances. Kin selection and reciprocal altruism are not, however, simple reflex systems that trigger invariable behaviors—rather, they are mediated by emotional effectors to produce a range of attitudes out of which behaviors are shaped.

Kin selection describes a genetic predisposition to cooperate freely with our relatives, the closer the relation the more spontaneous and extensive the cooperation tends to be. This mechanism expresses the logic that "blood is thicker than water." Kin selection became fixed in our genetic heritage because this mechanism encouraged behaviors for assisting in the survival of those individuals who (by their kinship to the helper) carried genes for helping relatives. Attentive parenting, for example, will assure the survival of children bearing genes for attentive parenting. And so much the same for generous "uncling," "cousining," and so on.

Reciprocal altruism, by contrast, is a genetically endowed predisposition for cooperating with nonkin under circumstances promising a high probability that favors given will be returned. "I'll scratch your back if you scratch mine" is the logic of reciprocal altruism. It is obvious that a tendency to risk alliances with nonkin would not become part of the genetic makeup in species unable to keep a running tally on the exchange of favors. But where these abilities exist (as they certainly do in humans, apes, dolphins, and many other species), traits for initiating conditional alliances and for retaliating against cheaters could become genetically fixed, for such traits would have significant survival value.

Kin selection and reciprocal altruism provide sufficient means for the kind of social organization typical of higher primates and early humans. Kin selection assures cooperation among close relatives, while reciprocal altrusim allows for alliances such as pair bonding and broadly extended families. These two mechanisms, operating together, constitute the genetic foundations for all particular forms of human social organization. It must be stressed once again that the dynamics of kin selection and reciprocal altruism are mediated by emotional factors such as affection, sympathy, gratitude, guilt, jealousy, resentment, and the like. In the absence of these emotional regulators there would be little hope for social coherence among humans. Parents having no affection for their children cannot be counted on to make even the minimal sacrifices required of them, and nonkin who feel no gratitude for favors received will be unlikely to reciprocate. But there is a further point involving the social utility of the emotional life: because human emotions lend themselves to manipulation by symbols it becomes possible to exploit them for the process of wising up to new forms of social organization.

In the infancy of our species we lived together in *kinship bands*, relatively small and stable groups of closely related hunter-gatherers. With a measure of good fortune kinship bands might well experience an increase in size together with a diffusion of genetic relatedness. Increased density of population would eventually place heavier demands on available resources, in which case the group would have to move more frequently or else hunker down until circumstances improved. If conditions did not improve then a group might show signs of stress along family lines. Conflicts of interest and outbreaks of hostility would become more frequent until part of the group took it upon themselves to organize and emigrate. This pattern of spreading out in response to environmental and social stress was the rule for most of human history. By this process of growth and fission, humans came to occupy every habitable continent of the planet well before the agricultural revolution.

The principal means of solidarity and cooperation in kinship bands were genetic, that is, the dynamics of kin selection and reciprocal altruism. These groups remained stable and coherent without having to articulate moral and legal precepts. As long as resources were in fair supply and as long as the group was of moderate size the genes would be sufficient to regulate appropriate attitudes and behaviors. The principal disadvantage of kinship bands was that the genetic means of solidarity and cooperation were all they had going for them. In good times these might be enough, but in times of crisis the genetic glue would be too weak to prevent social fission. This typically meant that kinship bands would remain small and mobile. While there is no inherent hardship in being small and mobile, consider what might befall a migrant kinship band that happened to stray into the territory of a larger, settled group. If, perchance, the larger group were flush with resources it might find a way to absorb the wandering band. But if (more likely) they were close to hunkering down they would probably deploy a fighting force to kill off the intruders.

But how did there get to *be* large settled groups to begin with? If the social mechanics of our ancestral environment enforced a pattern of fission among nomadic groups then by what steps might a group break out of the pattern? The answer is that larger, settled groups became possible by virtue of wising up to *cultural means* of solidarity and cooperation. Kinship bands tend to fission when stress outpulls the strength of the genetic bond. But if a group were able to invent some artificial bonding mechanisms to supplement the genetic factor, then social fission might be prevented. It is just such artificial means of solidarity and cooperation that we find emerging in *tribal alliances*. Tribal alliances were much larger groups dwelling in interactive villages, practicing light horticulture and herding to supplement forays of hunting and gathering. The enlarged size of tribal groups helped to defend against the ambitions of wandering bands, but also enabled the practice of raids against smaller, more vulnerable groups. Wising up to the ways of tribal organization was a good thing to do.

The secret of forming a successful tribal alliance was to find the articulate means to expand and reinforce the dynamics of kinship. Such means, if effective, could artificially strengthen emotional ties and thereby prevent the appearance of fissures that might result in the spread-out option. If the regulation of behavior could be transferred to an articulate (extragenetic) calculus then the means would be in place to expand solidarity and cooperation beyond the limits achievable by the genes alone.

This transference is precisely what the architects of tribal alliances were able to accomplish. By introducing a variety of symbols, rituals, and concepts it became possible to manipulate individuals into cooperating more freely with others. A new set of external emotional triggers was put into place, thereby overriding the influence of tacit signals. The result of these new cultural means was to redefine the boundaries of social reality, such that those who could be symbolically identified as belonging to the tribe were regarded as kinfolk, and were thus deserving of one's cooperation. In other words, symbolic markers had the power to create a virtual kinship, which is just as effective as the real thing. Among the decisive markers were body ornaments (e.g., scarring, tattooing), verbal greetings and gestures, distinctive styles of dress, and the like. In addition, there were communal events that served to reinforce the bonds between constituent family groups. Storytellers were especially important, for they could narrate events of the past in a way that objectified the common ancestry of all members of the group. They might even foretell events of the future, and in so doing they could formulate and legitimate shared goals and aspirations. Ceremonious gift exchanges were also important as means of establishing bonds of affection and gratitude, or perhaps for mending the odd tiff. The regular assembly of special interest groups (age groups, sex groups, crafters, etc.) would add further to the artificial means of solidarity and cooperation.

The ability of our species to wise up to articulate means of social organization is the one factor that most clearly distinguishes us from our simian cousins, the great apes. It is by the elaboration of these cultural means that we become human *(homo sapiens = man the wise).* The transition from kinship bands to tribal alliances was perhaps the most decisive event in the history of our species. It meant that cultural factors would henceforth interact extensively with genetic determinants of behavior. After this rubicon had been crossed the whole future of the species was to become a matter of deciding *which* articulate formulations of wisdom carried the most favorable odds for survival. Such matters are at the core of philosophical and political discourse.

Given adequate defenses against invasion, the tribal alliance was typically a very stable form of social organization. But if the combined impact of population and consumption placed too much stress on natural systems then a tribal group, like any other form of social organization, would be faced with the normal range of options. Perhaps the most common among tribal groups was the sit-back option. Let nature make the hard choices. When humans undertook the domestication of animals they invited upon themselves a spate of new diseases. Humans share sixty-five diseases with dogs, fifty with cattle, forty-five with goats and sheep, and forty-two with pigs. Smallpox, measles, tuberculosis, diphtheria, influenza, leprosy, and the common cold are just a few of the scourges that came along with the settled life. If a tribal group came to the point of exceeding the carrying capacity of its environment there was always a good chance that nature would step in with a timely epidemic.

Other options were common as well. When resources were in short supply a tribal group would be forced to hunker down for a spell. But if circumstances did not soon improve, the artificial means of solidarity and cooperation might begin to break down, resulting in the emigration of splinter groups. The kill-off option was common too. Infanticide probably increased along with population density, as it has been observed to do in many primate species. Another well-known phenomenon among tribal groups is ritual warfare, where periodic cycles of killing are triggered by environmental stress factors.

These patterns of death and dispersal kept human tribal groups in a state of fluctuating equilibrium with their environments. But at certain points along the way some tribal groups took the option of wising up to new means of solidarity and cooperation. Thus the *chiefdom* emerged as a new form of social organization.

Wisdom in a chiefdom was, in essence, the wisdom of the chief. Whereas tribal groups emphasized consensus and decentralized decision making, chiefdoms focused all authority on the figure of an absolute ruler. Chieftains ruled by a combination of charisma and respect, backed by military power and dynastic ideology. The chief commanded virtually all the symbolic and material resources of the group, enabling him to manipulate at will the emotional effectors (and thus the behavior) of his subjects. All members of the group affirmed kinship to one another by their common descent (usually contrived) from the chief's clan. And the dominance of the chief himself was typically legitimated by the myth of divine favor. Cooperation within the group was assured by the logic of dependence and veneration—the chief was perceived with awe, a majestic and generous figure whose wishes brought willful compliance from individuals throughout the realm. Failing this, the chief might inspire fear by virtue of his sovereign power. It made good sense for tribal alliances to wise up to the totalitarian rule of a chiefdom because such centrally administered groups were far more efficient at producing and distributing goods, and they were formidable military forces in a context of increasing hostilities between tribal groups.

Chiefdoms are notoriously unstable social systems, becoming increasingly vulnerable to collapse when brought under stress by dwindling resources. As a chiefdom approached ecological limits it would be forced to hunker down, and typically the resulting sacrifices would be unevenly felt, reflecting the fabric of the chief's power base. Continued hardship would increase the likelihood of political unrest. To prevent revolt a chief might attempt to relieve stress by culling out the realm, for example, by declaring an increase in infanticide, or by selectively wiping out entire villages. More common, though, would be the spread-out option, which typically meant wars of conquest and colonization. Therefore, the basic pattern of chiefdoms tended to be growth, not equilibrium. Too much growth, however, would eventually strain the administrative abilities of the chief, leaving him too thinly spread among his duties to be effective at any of them. Whether by reason of neglect, incompetence, unfairness, or sheer bloody-minded cruelty, overgrown chiefdoms tended to become cauldrons of inequity, seething with discontent and intrigue, inching toward the boil of political chaos.

The political chaos of a failed chiefdom might finally be resolved by wising up to a new form of social organization, the *state system*. States arise as the various functions of the chief are taken over by bureaucratic institutions, thus transforming the social order from a centralized system dominated by personal whim to a hierarchical system regulated by formal policy.

Above all other considerations the state system must be seen for what it does to transform the means of solidarity and cooperation. Official symbolism and ideology typically displaced ancestral relations to the chief as the principal means of social solidarity. Thus do states take measures to assure that citizens internalize a common set of religious and political ideals, the price of which has been paid by epic sacrifice, and the promise of which is projected in a vision of future security and prosperity. The citizens of a state are thus emotionally galvanized by shared history, commitments, and hope. These means of solidarity are constantly reinforced by a variety of public symbols and ceremonious events. States offer a variety of institutional means for cooperation. These include monetary and market systems for the efficient distribution of goods and services. But perhaps the most important functions of the state are in maintaining political and legal systems, the former to select leaders and the latter to legislate, enforce, and adjudicate formal laws.

I said earlier that the most distinguishing feature of our species is its ability to wise up, that is, to invent cultural means for reinforcing or overriding genetically predisposed behaviors. This ability has enabled our species to transcend the conditions of simian existence and to develop radically new patterns of association and cooperation. When we are pushed to it, that is. When stress factors appear in natural and social systems as a result of overpopulation and overconsumption, then established patterns of solidarity and cooperation begin to fail, leaving social groups with a limited range of options for relieving stress. A fundamental thesis of this book is that the elements of the contemporary global problematique have pushed our entire species to the point where it is clear that existing means of solidarity and cooperation are incommensurate with the challenges we face. It is time to wise up on a global scale.

Wising Up

The challenge of wising up in the twenty-first century can be easily stated, though it will be a task of humbling complexity to achieve. There are really two principal goals, one social and the other psychological. The social goal is to reharness certain features of our biological heritage and place them in service to a larger vision of solidarity and cooperation. We have seen that humans are genetically predisposed to associate and cooperate with kinfolk, and to make mutually beneficial alliances with nonkin. We have further seen that these dynamics (i.e., kin selection and reciprocal altruism) are regulated by a repertoire of emotions. And these emotional regulators have proven themselves to be open to manipulation. In other words, by modifying key symbols it becomes possible to redirect the emotional responses (and therefore the behaviors) of individuals. The secret to wising up in response to the global problematique will be to come up with the right symbols that will engage the emotions in a new way, such that the result of our doing so will be to enhance global solidarity and cooperation. To simplify, our task is yet again to enlarge the tribe, that is, to expand by symbolic means the range of our affection, sympathy, gratitude, and guilt to include all members of our species, even those of future generations. In the powerful way that national symbols can create a feeling of oneness among diverse compatriots, so we need the symbols that can work the same magic on a global scale. Only then can we begin to think and to act as a single species.

The psychological goal is similar, though it interferes with the ways we see ourselves rather than the ways we see others. The global problematique has reached its present level of crisis because of excessive patterns of population and material consumption. The behaviors driving these patterns are, in turn, linked to individual needs for self-esteem. In many overpopulated countries of the southern hemisphere a large family is regarded as a measure of personal worth. And in the industrialized countries of the northern hemisphere individuals are conditioned to associate personal worth with the acquisition of material possessions. Thus has the need for self-esteem functioned as a major contributing factor to the crisis. Yet there is nothing essential about these links between self-esteem and the behaviors resulting in overpopulation and overconsumption. Self-esteem is a variable phenomenon and may be linked to virtually any type of behavior. Some people may be motivated to seek self-esteem by acquiring material possessions, while others may acquire a sense of self-worth from artistic achievement or by performing charitable acts. Wising up in response to the global problematique calls us to decouple self-esteem from destructive behaviors, and instead to link the achievement of self-esteem to behaviors that enhance the integrity of natural and social systems. This task calls for a transformation of values at a very fundamental level. Again, this task is related in many ways to the symbolic life of a culture.

Together, these two tasks call us to participate in a transformation of social and psychological realities on a global scale. The complexities of this mission boggle the mind, for our understanding of the dynamics of social and psychological transformation is limited. And what is worse, we are working against a time factor that is, to say the least, very discouraging. A transformation of social reality will involve the creation of new institutions and organizations at all levels, as well as the demise of others. In particular, the fate of the nation state is brought to point. The nation state currently functions as the ultimate unit of solidarity and cooperation. Yet the nation state is proving its obsolescence by the way it obstructs the level of solidarity and cooperation commensurate with the global problematique. Present demands for broader cooperation suggest movement toward a global parliament, while demands for more effective problem solving at local levels suggest measures toward decentralization. National governments may be expected to undergo a process of redefinition and rescaling as they respond to these divergent pressures.

A transformation of psychological reality would amount to redefining the standards for judging self-worth. As I said earlier, self-esteem can be linked to an unlimited range of values. Not all of these, however, are consistent with undoing the global problematique. What is called for is a moral calculus that will motivate individuals to act in ways that will reduce population and material consumption. To this end there are a few relevant facts that can be asserted with confidence. First, we know that a certain minimal level of economic prosperity is necessary for self-fulfillment. Second, we know that no amount of prosperity in excess of this minimal level will be sufficient to guarantee self-fulfillment. And finally, we know that elevating the economic status of impoverished women results in a lowering of birth rates. Given these considerations there appears to be no reasonable alternative to the view that wising up in response to the global problematique calls us to enhance the economic prosperity of impoverished regions of the world while simultaneously reducing the prosperity of affluent regions. There is no good reason to believe that such a program of global economic leveling would be possible without strong moral support from individuals the world over. And nothing short of a radical transformation of moral conscience could provide this support.

Clearly the transformation of social and psychological realities must go hand in hand. Individuals cannot be expected to make radical changes in their values when they are immersed in social contexts that reinforce the status quo. Nor can we expect new institutions and organizations to emerge without the moral leadership of individuals inspired by new values. So how might we proceed to address this monumental task of reconstructing basic social and psychological realities? We only need to inquire about the sources of the social and psychological realities we presently have. That is, where do we look to find those symbolic means by which humans have always acquired their most fundamental visions of personal and social reality? We look to their stories.

Stories Are Us

So far the argument has claimed that human folly has created the conditions for global, systematic, immediate, and chronic threats to the integrity of natural and social systems, and that the only adequate response to these threats is to wise up to new means for global solidarity and cooperation. I have further suggested that wising up in the twenty-first century calls for a reconstruction of social and psychological realities, that is, a reorientation of collective and personal self-understanding. The argument now continues with the claim that the most fundamental sources of self-understanding are found at the level of story. To put the thesis even more directly: if we are to respond effectively to the global problematique then we must attend to our stories, for in the story of a culture we find its most profound expression of wisdom.

Perhaps this thesis exaggerates the case. Is it possible that stories can make a decisive difference in the viability of our species? Is story *that* fundamental to human existence? It is difficult to believe otherwise once we stop to consider the many ways in which our lives are shaped by stories. We legitimate institutions and values in their name, we wage wars in their defense, we judge ourselves and others by their standards, we take pains that our children will learn them well, we draw inspiration from their examples, we construct our hopes and fears under their influence, and so on. It would not be extreme to say that we negotiate our way through life by the guidance of our stories. We humans are the only species to tell stories, and so far as anyone knows every recognizable human community has fostered storytelling traditions. Thus we appear to be in the presence of a trait that is both universally and exclusively human, suggesting that what humans *are* has something to do with their stories. If these ideas hold up—that is, if storytelling is an essential human activity bestowing substance and form on the lives we have—then it would follow that changes at this level of human thought will be among the most profound and far-reaching we can imagine. Thus we have good reason to believe that appropriate changes at the level of story might hold the power of reorientation needed for enhancing human solidarity and cooperation. Such is the fundamental premise of this book.

For this premise to bear the weight of further discussion it will be important to be clear about what is meant by story. Many different things fall under the rubric of story, most of which are unworthy of the rather large claims being made here. Stories may be of many different types and are told for a variety of reasons. They may be true, fanciful or fictitious, and they may be told to entertain, to instruct, to exhort, to edify, to explain, to excite, and so on. If the many forms of story have anything in common it is only that they relate a series of connected events. So what sense of the term do I have in mind when I say that story gives substance and form to human life? I have in mind the *profound* sense of the term. And what might *that* be? In the profound sense a story is intended to be true rather than fictitious or fanciful. That is, it offers a narrative of connected events that is meant to express an adequate account of the way things are in the real world. The profound sense also suggests that the information contained in the narrative is important or valuable for the receiver's interests. Thus, in the profound sense, story relates a series of connected events that are both real and deeply significant.

But I intend even more than this. I want to suggest that the profound sense of story bears a deep relation to the basic functions of the central nervous system. I am saying that if we can picture what the brain does for an individual organism then we shall have a way to think about what story does for a cultural tradition. All brains (human or otherwise) do essentially the same thing, that is, they take some measure of how things are in the external environment and then use this information to devise behaviors that will be more or less conducive to the interests of the organism. Brains that have become specialized for symbolic interaction do not suddenly depart from this basic function, they just continue it in a new domain, the domain of culture. Indeed, human culture *just is* the result of assessing and addressing the environment with the aid of symbols. In every particular human culture, therefore, we may expect to find (and *do* find) two basic types of ideas: ideas about *how things are* in the world, and ideas about *which things matter* for human existence. This principle may be extended by the assertion that a culture exists as a coherent entity to the extent that its members share common ideas about how things are and which things matter. A further implication is that the important differences between cultures may be measured in terms of incompatible ideas about reality and value.

This is where the profound sense of story becomes essential to the life of a culture. It is obvious that an absolute consensus about how things are and which things matter could never be achieved, not even in the simplest of social groups. You cannot get everyone to agree about everything. But an absolute consensus is not necessary for ample solidarity and cooperation to occur. It is sufficient that there be a broad consensus regarding a central core of ideas, that is, those ideas that articulate how things *ultimately* are and which things *ultimately* matter.

It may be useful to construct a model to help us in thinking about the nature of human culture (see Figure 1, page 24).

At the core of every cultural tradition there is a narrative, a myth, which integrates ideas about reality and value. The narrative core provides the members of a culture with certain information that gives them a general orientation in nature and in history. The narrative core is the most fundamental expression of wisdom in a culture—it tells us about the kind of world we live in, what sorts of things are real and unreal, where we came from, what our true nature is, and how we fit into the larger scheme of things. These are all *cosmological* ideas, they inform us about the cosmos and our place in it. But the narrative core also contains ideas about *morality,* that is, it tells us what is good for us and how we should conduct ourselves in order to achieve our fulfillment. In the narrative core of a cultural tradition facts and values are interwoven in a seamless series of connected events, in precisely the way that cognitive and emotional events are integrated in the life of an individual.

Narratives integrating ideas about how things ultimately are and which things ultimately matter function as important resources for the construction of personal and social realities. In other words, it is by virtue of story that the transsimian phenomena of personality and culture begin to appear. Stories facilitate transsimian existence by establishing new possibilities for social organization and personal fulfillment. Without them, humans would soon default to the psychological and social circumstances of the great apes.

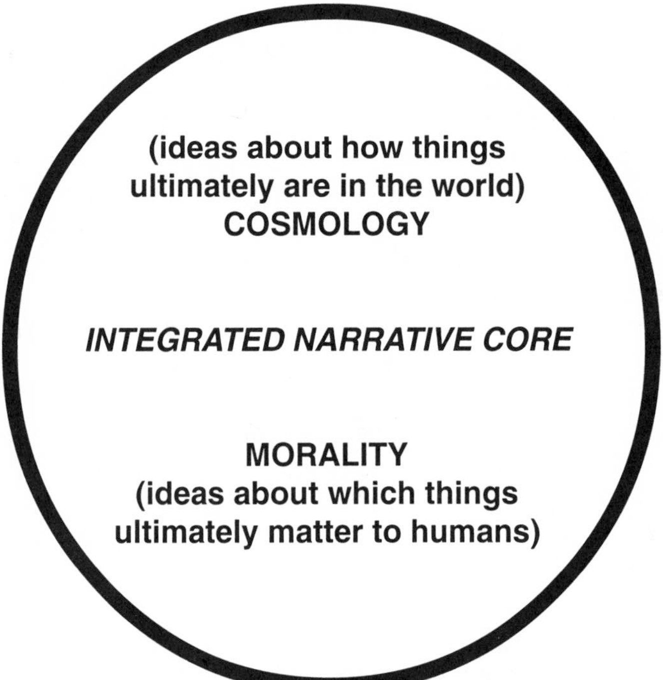

(ideas about how things
ultimately are in the world)
COSMOLOGY

INTEGRATED NARRATIVE CORE

MORALITY
(ideas about which things
ultimately matter to humans)

Figure 1

Individual homo sapiens become persons when they become moral agents, that is, when they acquire socially induced motives and standards for judging self-worth. In large measure, individuals acquire these elements of personal identity by ingesting the general features of a narrative core. Likewise, an accidental collection of individuals may become a distinctive and coherent culture by constructing a consensus of ideas among individuals. As persons assimilate a narrative core of ideas they acquire a common worldview, common values, common standards of rationality—in short, they acquire new resources for solidarity and cooperation. The ultimate measures by which social progress and personal success are judged have their moorings in the story of a culture. Stories are us.

It is probably not the case that new forms of social and personal existence were carefully worked out in long-range planning committees and then implemented after a general referendum. It is more likely that groups were forced to come up with ad hoc solutions and strategies in the chaos of local crises. Stories would play a decisive role in this process because they had the power to stabilize patterns of behavior by elevating them from the status of accident and ad hockery to the status of legitimate and normative practice. Stories performed the function of showing how the newfound lifeways of a culture were an integral part of a larger, more ultimate scheme of meaning. They have the power to catch serendipity on the wing and to lock it into normative traditions that then define the possibilities of human life. Narrative accounts of how things ultimately are and which things ultimately matter found expression in the various symbolic vocabularies of ancient myths, art, religions, and philosophies, where their influence was to structure the intellectual and moral lives of individuals.

I have suggested that without effective story traditions humans would soon default to the social and psychological circumstances of the great apes. As if to preclude this possibility of backsliding, various *ancillary strategies* have developed within story traditions to maintain a high degree of consensus about the narrative core. Thus a mature cultural tradition may be seen to have a characteristic structure, with an integrated narrative at the core and various supportive strategies at the outside, as shown in Figure 2 (see page 27).

A brief explanatory note on each of these ancillary components of story traditions will be sufficient.

Intellectual The intellectual domain is normally tended by philosophy and/or theology. These disciplines help to maintain the integrity of the story by clarifying its meaning. They also help to defend the plausibility of the story against both internal and external critics.

Institutional Most story traditions have an institutional dimension that oversees many of the social aspects. These would include the administrative task of regulating the transmission (i.e., education, propagation) of the story, together with the development of policies for resolving conflicts arising within the group.

Aesthetic Most individuals are drawn toward tangible expressions of the narrative core. Thus most traditions have developed characteristic art forms (e.g., painting, sculpture, architecture, music, and poetry) to give the narrative an objective presence.

Experiential In many story traditions individuals are encouraged to seek some sort of profound religious experience. Such experiences are important because they typically provide personal validation for religious beliefs.

Ritual Virtually every story tradition has ritual practices that make it possible to relate the particular events of life to the narrative core. The point of ritual is to make a connection between the events of daily life (positive and negative, routine and extraordinary, personal and communal) and the timeless events of the story. In this way one apprehends the universal significance of everyday life. When individuals are able to make these connections they take ownership of the story.

Once again, the point of these various components of story traditions is to establish and maintain a broad cultural consensus

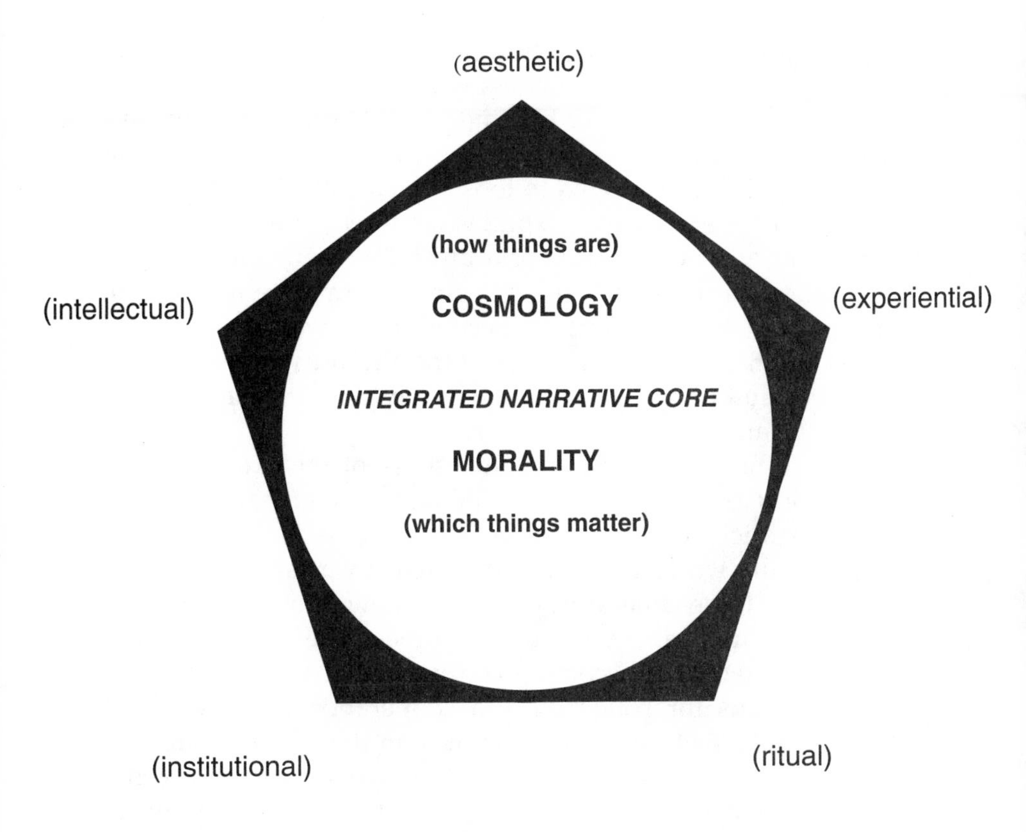
(aesthetic)
(intellectual)
(experiential)
(how things are)
COSMOLOGY
INTEGRATED NARRATIVE CORE
MORALITY
(which things matter)
(institutional)
(ritual)

Figure 2

regarding the features of the narrative core. When these strategies are effective there will be little question about the intellectual plausibility and the moral relevance of the story. While the influence of these various ancillary strategies helps to safeguard a cultural tradition from defaulting into simian existence, they make it equally difficult to effect radical progressive changes. Minor adjustments may be allowed, but anything approaching a transformation of consciousness will encounter suspicion and resistance. Radical changes become possible only when the ancillary mechanisms lose their effectiveness and the prevailing wisdom of a culture begins to look implausible and irrelevant. When this happens we may expect to see three general responses. First, there will be evidence of a breakdown of solidarity and cooperation as the social order begins to gravitate toward the conditions of simian existence. Second, there will emerge a variety of militant attempts to reassert the authority of the prevailing story tradition. And finally, new voices will be heard offering a range of alternative stories.

We are now coming to a better sense of the problems and the prospects for wising up in response to the global challenge. There is sufficient evidence to believe that the major story traditions of the world are losing their effectiveness, with the result that the three general responses are now upon us. These are the best of times and the worst of times. The worst of times because the world is becoming confused and dangerous as the cultural means for solidarity and cooperation slip away. But these are also the best of times because in these chaotic circumstances there appears a chance for a new world order to ascend, to be called forth by a compelling vision of how things really are and which things really matter.

Axial Periods

The central message of this book is that everybody's problem calls for everybody's story. That is, a set of global, systematic, immediate, and chronic challenges calls for responses that are likewise global, systematic, immediate, and enduring. In order to initiate the many specific projects and programs that must be undertaken worldwide we must learn to think and to act as one species, unified by effective means for global solidarity and cooperation. Such means, I have argued, are the province of story. An adequate response to the global problematique entails, therefore, the articulation of a vision of reality and value that is sufficiently compelling to elicit a genuinely global culture.

But is it even remotely possible that a new story tradition might arise to transform social and psychological realities on a global scale? Terms like "new world order" and "global consciousness" roll off the tongue with ease, but can one seriously propose what they intend without revealing a deep naïveté about social, psychological, and historical dynamics? The emergence of a universal wisdom tradition rich enough in means to give birth to a global culture would be an event without precedent in human history. And yet such an unlikely event begins to look believable as one examines the circumstances in which today's dominant story traditions first appeared.

In the period roughly surrounding the sixth century B.C., in various parts of the ancient world, there occurred a remarkable coincidence of events that transformed the social and psychological organization of the human species. During this period, the so-called Axial Age, there appeared a wave of revolutionary prophets, teachers and poets whose influence has carried down to the present day. In China this was the period of Confucius and the legendary Lao-Tse, the period when all the major schools of Chinese philosophy were founded. In India it was the time when Buddhism and Jainism originated, and when Hindu traditions were being reformulated by the writing of the Upanishads. In Persia Zoroastrianism appeared. In Israel the revolutionary prophets and the new priestly class were laying new foundations for Judaism, the tradition that would eventually spawn both Christianity and Islam. And in Greece the Orphic tradition was transforming Greek religion, while a golden age was dawning for philosophical speculation and dramatic poetry. The Axial Age exploded with new forms of wisdom.

The diversity of these religious and philosophical movements was considerable, and speaks against the hypothesis that they had their origins in a single source, or that there was extensive cross fertilization among them. All of which makes the deep formal similarities of these movements the more impressive. The prophets and teachers of the Axial Age were telling a new kind of story about how things are and which things matter, and the stories of these traditions continue to influence the intellectual and moral lives of the vast majority of individuals alive today. While this is not the place to describe the Axial Age at length and in detail, there are four general features of the period that have relevance for the discussion under way.

1. *A sense of intellectual and moral failure.* The centuries leading up to the Axial Age were a period of increasing complexity and confusion. Populations were increasing dramatically in each of the axial regions, placing natural and social systems under stress and triggering the options for warfare and mass migrations. Much of the population increase was absorbed by cities. Rapid urbanization will generate problems under the best of circumstances, but the problems multiply when the new urban dwellers hale from diverse backgrounds, as was the case. This was a period of experimentation with the organs of the state system, which meant that social circumstances would be complicated by false starts amid spasms of residual tribalism and despotism. Diverse peoples were drawn or forced into the proximity of growing cities, there to discover and contend with their differences. In general, the pre-Axial Age was a period of rapid growth and migration, where turmoil was punctuated by conflict and warfare. Time to wise up.

Clearly, the available cultural means for solidarity and cooperation were not up to the challenge of the new social pluralism. The prophets and teachers of the Axial Age interpreted the chaos of the period in their own distinctive ways, but the underlying themes were the same: there has been a breakdown of moral order, the present generation is degenerate and bereft of wisdom, fundamental change is necessary.

2. *A new vision of cosmic order*. Axial thinkers construed the chaos of their circumstances in terms of a problematic discrepancy between actual and ideal states of affairs. It was a matter of wisdom; humans were not living in harmony with reality. But if chaos resulted from moral failure, then order and harmony could be restored by moral rectitude. These perceived links between chaos and actual behavior, on the one hand, and between order and righteous behavior, on the other, could not be merely accidental. There must be some sort of objective reality involved, some independent moral order that has been transgressed. Such insights may have been implicit in the moral systems of pre-Axial cultures, but they were not fully worked out. With the emergence of Axial traditions they became explicit and central. Now the discrepancy between actual and ideal states of worldly affairs could be viewed in terms of radical discontinuity and tension between the mundane order and a transcendent moral order.

It was common among pre-Axial traditions to posit the existence of a transmundane order, often the spirit world or the realm of the dead. In these traditions the "other" world was typically held to be a faint copy or shadow of the mundane world. The two realms were roughly parallel in all their details—distinct, but fundamentally similar and harmonious. In the cosmologies of the Axial traditions, by contrast, there was a strong tendency to sharpen the distinction between this worldly and otherworldly domains, to place them in opposition. Together with this sharpened distinction there came a tendency toward dualisms, contrasting divine and human, good and evil, sacred and secular. Each of the Axial traditions worked out the details of transcendental cosmology against the backdrop of their particular cultural heritage, but there remains a common theme consistently stressing the objective and independent existence of a higher moral order in tension with the mundane order.

As the cosmological features of the Axial traditions became more explicit so did the moral features. Implicit in the new cosmologies were moral imperatives that soon became elements within various schemes for salvation. The common theme of the Axial story traditions was that the transcendent moral order was governed by an eternal principle (e.g., law, reason, divine will), which was articulated in the form of moral wisdom. Salvation was to be achieved by bringing the mundane order into harmony with the transcendent moral order. Ideas about death came in for revision too. Most pre-Axial traditions held beliefs about human immortality, but these beliefs were not linked to moral behavior. Destiny in the world beyond was what it was, irrespective of how one had lived. But the Axial stories tended to moralize these matters, so that the quality of the life beyond became directly linked to one's moral character in the mundane order.

3. *An emphasis on individual morality.* Perhaps the most consequential achievement of the Axial traditions was a thorough individuation of the moral life. In the pre-Axial world there was virtually no systematic attention given to the development of moral character at the level of individuals. In tribal contexts moral judgment was a social phenomenon, where the moral life generally meant waiting for a consensus to descend on the group. In chiefdoms moral issues were the province of elites, leaving common folk, as children, to comply with direct orders from above. For the most part, pre-Axial societies had few expectations about the abilities of common women and men to engage in moral deliberation or to bear responsibility for moral acts. The sense that virtue could be its own reward was undeveloped.

The dynamics of morality were different in Axial societies where obligations, rewards, and punishments were radically personalized. In Axial cosmologies there was a tendency toward dualism, a categorical separation of realities that transcendentalized the moral order. A corresponding anthropological dualism detached the self as a unit from both the social and biological aspects of one's being. A new sense of the self, a soul or mind, emerged as the true identity of a person. This soul was an independent entity that could and should be changed, that is, it could be cultivated, worked on, refined, purified, or released by various measures of discipline, reason, piety, study, ritual, and so on. If there was chaos afoot in the mundane order it resulted from the failure of individuals to reconstruct their inner selves to conform to the moral principles inherent in the transcendent order. What appeared in the Axial Age was a diversity of speculative religious and philosophical movements offering resources for enriching the inner life. By these means each individual would have access to ultimate wisdom. Individuals acquired a sense that what transpired within their own souls would have important consequences beyond their immediate circumstances. Salvation had become radically personalized—a possibility for each individual, and no longer a vague quality of the group or the exclusive destiny of elite rulers.

4. *An emphasis on universal solidarity and cooperation.* The social implications of the Axial Age movements were no less profound than the psychological ones. The prophets and teachers of the new story traditions made universal claims for their wisdom. The deities, principles, and values of the new age were pronounced with cosmic significance, transcending ethnic and national pluralism to include everyone. These stories were intended for the entire species. Thus a new ideal of universal solidarity and cooperation came to replace the provincialism and exclusivism of the pre-Axial past. The oneness of humanity, though never completely realized, was clearly conceived by the new story traditions. And so was the piecemeal strategy for its achievement: conflicts could be resolved and the world could be unified by the transformation of individual souls. To this end new forms of social institutions emerged with a universal mission. Schools, churches, and monasteries appeared everywhere, organized to transform the social order by gentle force of spirit. By individualizing moral sensibilities and by universalizing moral ideals, these Axial traditions, powered by the emotional appeal of imaginative stories, negotiated a passage of the species to new possibilities for solidarity and cooperation.

The Axial traditions did not succeed in giving birth to a global culture, but they were able to expand human solidarity and cooperation on a scale that was unprecedented, even unimaginable, prior to their appearance. They managed to pull off the most impressive wising-up operation in human history. Our calling is no less than to achieve for our time what these ancient traditions did for theirs, that is, to transform social and psychological realities in ways that effectively redress the global problematique. To do this we must find the courage to be no less radical in our storytelling than were the Axial prophets and poets. Their achievement is our source of courage and hope for a new Axial Age.

Of course many will contend that the original Axial Age has yet to run its course. That is, perhaps it is premature to reject the possibility that one of the great religions of the world might ultimately succeed in realizing its universal mission. Why should it be assumed that we are called to wise up to a new story tradition when we already possess the final truth about how things are and which things matter?

No Story Is Final

In the final sections of this introduction I will venture a two-part thesis. First, there are good reasons to believe that the story traditions of the Axial religions have played themselves out, and that their influence will continue to decline. And second, we now see emerging the elements of a new story having the potential of becoming a narrative core for a global cultural orientation.

The Axial story traditions had a beginning and a middle, and it is reasonable to suppose they will have an eventual end. Their influence will probably continue to decline because for many individuals they no longer give a compelling account of how things are and which things matter. That is, their ancient cosmological ideas appear implausible to the imagination of the twenty-first century, and their moral vision appears irrelevant to the challenges of the global problematique.

Increasingly, since the Enlightenment, it has become clear that the stories of the Axial traditions are contingent caricatures of reality. They are the imaginative constructions of ancient cultures, developed in response to particular challenges of the distant past and expressed in the concepts and categories of primitive cosmologies. In their classical formulations the relativity and the inaccuracies of these stories are palpable. And to the extent that any account of reality is perceived as a contingent caricature its plausibility is compromised.

The contingency of Axial stories becomes evident as one becomes acquainted with the details of their historical development. Many examples could be cited here, but I will make do with a single one drawn from early Christian history. The earliest followers of Jesus were Palestinian Jews whose devotion to Jesus was based on his rendition of the Jewish Law. The authority of Jesus was initially a moral authority. As the Christian mission spread into Hellenistic territories it became evident that audiences were less interested in his moral authority than in his metaphysical credentials. In response to audience demand the Christian mission introduced the idea that Jesus was a cosmic figure, the Son of God, a being with authoritative metaphysical status. Thus the claim that Jesus was the Son of God is historically contingent on the ambitions of the early Christian mission. It must be granted that this account does not constitute a proof that Jesus was not a divine being, but it does place the claim in a context where it is most plausibly construed as a mere period piece.

The contingency of Axial stories becomes further evident as they are brought out of isolation and into proximity with other stories. When a person is familiar with a single narrative account of how things are and which things matter, this account will, by default, have the appearance of objectivity. But as one becomes familiar with two or three such accounts (as happens in a pluralistic age such as ours), then they take the appearance of equally viable alternative options, with no more than subjective criteria to recommend one over another.

The plausibility of Axial stories has also been impaired by the advance of modern science. When contemporary men and women are faced with questions about human origins, human nature, and human fulfillment they are far more likely to accept the accounts offered by the sciences than those offered by traditional mythologies. Before the rise of modern science it was possible to accept mythological stories in a literal and straightforward manner. But with the advance of the sciences such accounts came to be viewed as grossly inaccurate descriptions of how things really are. In a literal sense they became straightforwardly false. In order for the modern mind to make sense of ancient mythologies it must interpret them through the lens of scientific concepts and categories. But the strategy of reinterpretation always compromises the distinctiveness of the original stories. When St. Augustine used neo-Platonic categories to reinterpret Christianity he created a worldview that was fundamentally neo-Platonic and incidentally Christian. The plausibility of Augustinian Christianity was made contingent on the plausibility of neo-Platonism. The same thing happened when St. Thomas produced an Aristotelian interpretation of Christian myth, that is, he reduced the meaning of Christianity to the substance of Aristotelian philosophy. The old wine always derives its taste from the new wineskin. The rise of modern science has therefore placed the Axial myths in a precarious position—to the extent that they can be rendered plausible to the modern mind they will have sacrificed their distinctiveness to the more fundamental meanings derived from the scientific worldview.

My point here has been that the Axial story traditions are in a state of declining influence in part because their recognition as contingent caricatures has rendered their accounts less plausible to the modern imagination. Of course it may be argued that *every* account of the way things are can be shown to be a contingent caricature, even scientific accounts. Recent work in the philosophy of science has severely compromised the view that scientific knowledge is completely objective and undistorted by bias. However this may be, it remains that judgments about plausibility are comparative judgments—that is, any account of how things are might be judged more or less objective and accurate than another. But at the end of the day scientific accounts will appear the most plausible for the simple reason that the methods of science represent a systematic effort to minimize contingency and distortion.

The influence of Axial traditions will continue to decline as it becomes ever more apparent that their resources are incommensurate with the moral challenges of the global problematique. In particular, to the extent that these traditions have stressed cosmological dualism and individual salvation we may say they have encouraged an attitude of indifference toward the integrity of natural and social systems. At the very least we have to admit that these story traditions did not prevent the behaviors that produced the global problematique. After all, the crisis developed under their watch.

Dualism and individualism have the effect of relativizing the mundane order of natural and social systems. They encourage the view that what is most essential about human existence transcends nature and society, that the physical, biological and social aspects of our being are of negligible value. What *really* matters is the spiritual aspect. Thus humans are entreated to rise above embodiment and community to seek fulfillment in the transnatural order of being. This is not to say that Axial stories disparage nature and community as something evil, but rather that they relegate them to the nonessential.

Perhaps, given the circumstances peculiar to the Axial mind, dualism and individualism were merely the most obvious or appropriate ways to symbolize human uniqueness and moral responsibility. Even so, there has been a moral price to pay for these symbols. The consequence has been to allow humans to think of themselves as essentially *apart* from the mundane order, thereby free to overlook the integrity of natural systems in the pursuit of ideals, goals, and projects. Precisely the problem.

Indeed, humans *are* unique, but our uniqueness in no way removes us from the realities of nature and community. We are emphatically natural and social beings, whose dependence on biological and social systems is absolute. It is too much to suggest that *anything* human might be transcendent. No biological niche, no species; no social matrix, no personality. However badly we want to affirm human uniqueness it cannot be done in ways that remove us from the conditions of our unique existence.

We see this now, of course—and we see it with such clarity and conviction that any failure to see it will appear to be out of touch with reality. Any story claiming moral relevance to the global challenge will not only factor in the mundane conditions of our existence, but will go beyond this to proclaim them sacred.

Despite the impressive record of the Axial story traditions one must ask about their ultimate failure to realize the ideal of universal solidarity and cooperation they have so persistently espoused. Such a failure is yet another issue bringing to point the relevance of these traditions. In fact, the record is mixed. These traditions have been as much for war as for peace, as likely to splinter into parties as to build consensus. But why is this so? Why diversity and hostility when the whole point of these stories is to create the conditions for unity and peace? In this case the obvious answer is probably the best, that is, none of these stories has proven sufficiently compelling to command a universal audience. Each of them lacks the objectivity it would take to make the competition—whether an internal variant or an external alternative—look implausible or irrelevant. Thus the failure to prevent diversity. And lacking in these resources their only consistent defense against a hemorrhage of diversity has been to assume a posture of dogmatic authority. Hence the hostility.

It finally appears that the major story traditions of the world have taken us as far toward global solidarity and cooperation as their resources will allow. To take us any further they would have to transcend the limits of their appeal. But as things stand there is little prospect of this happening. Can Islam make Christianity look implausible to a committed Christian? Does Judaism hold much appeal for a Buddhist or a Hindu? To the adherents of these traditions the others will continue to look like someone *else's* stories, none of them compelling enough to displace their own. The Axial legacy, it seems, has brought us to an impasse.

I have been suggesting that we may expect the major story traditions of the world to decline in their influence as satisfying ways to organize social and psychological realities. Developments of the past few centuries have brought each of these traditions to a double crisis of intellectual plausibility and moral relevance. Many of the cosmological claims of traditional stories have been rendered either untrue or unlikely by developments in modern science, leaving them either impossible or difficult for contemporary women and men to believe. And further, the failure of traditional stories to transcend cultural barriers, and their palpable lack of resources for addressing the underlying forces of the global problematique make them appear out of touch with the real world. The crisis of plausibility has rendered our traditional stories unsatisfying in their accounts of how things are, and the crisis of relevance has rendered them inadequate in their judgments about which things matter. These concerns indicate a significant need, both psychological and social, for a new story to provide the means for a transformation of the species on a global scale.

Having said all this, I must now soften the foregoing critique of existing wisdom traditions by conceding that I may be completely mistaken about their future. My remarks have assumed an identity of these traditions with their classical formulations. That is, when I suggest that the Axial traditions appear implausible and irrelevant (and thus destined to decline), I am speaking of the versions of these traditions that one finds in conventional and popular piety. But of course these are *living traditions*, and as such they may still find the resources and the will to change in fundamental ways as they absorb the perspective of modern science and feel the impact of the global problematique. There may be good reason to believe that the Axial traditions, sufficiently transformed, will be able to play a constructive role in the birthing of a new story. Concerning these matters, more later.

Now for a New Story

When the wisdom of the past appears implausible and irrelevant, when the stories on offer appear as so many contingent caricatures, then individuals begin to hanker for a new story. On the strength of this premise I have been proposing a quest for a new wisdom tradition, a new story, everybody's story, that might give us the intellectual and moral resources to address our urgent problems.

The obvious questions to emerge in response to such unfettered optimism are these: Is such a story desirable? and, Is such a story feasible? Unless we can give affirmative answers to these skeptical challenges the quest for everybody's story is off.

Is everybody's story desirable? An informal survey of cultural trends around the world suggests just the opposite is the case. There is a lot of intense interest in the topic of narrative these days, but most of it appears to be fixed at the level of particular stories, not at the level of a universal one. Indeed, interest in particular stories has reached the proportions of a global movement, variously known as multiculturalism, cultural diversity, or pluralism. Multiculturalism has become a serious movement in various parts of the world because it appears to free individuals and minority groups to define their own identities and to influence their own destinies by formulating their own particular experiences. The mood inspiring this movement says that all particular groups of people have an inalienable right to be known by their own stories and not by the stories imposed upon them by dominant groups. Multiculturalism amounts to a democratization of story, and it is both psychologically and politically salutary. It says, "we are valuable, we are somebody too, and *we* get to say what is good for us—so listen to our story." This is a constructive attitude.

Yet there is no reason to think that the value of formulating one's particular story in any way displaces the need for a universal story. In fact, the need for a universal story is amplified by the experience of recovering particular stories. It is true that to know ourselves is to know our particular stories, and to affirm ourselves is to tell these stories. But the more we come to know and to appreciate them the more we become aware of their particularity. A particular story may be mine, and it may be worthwhile, and I may be diminished without it, but it is not a story that speaks for everyone's experience. And as I discover the limitations of my own story there is born within me a longing to hear the larger story of which my own is a part—the universal story, everybody's story.

But what about the feasibility of telling everybody's story? Do we have the resources for it? The multicultural movement has helped us immensely in providing resources for exploring and affirming our particular stories. There are books and workshops on coming to terms with sexuality and gender, there are genealogy centers and databases to help us rediscover our ethnic roots, we have oral history projects and historical societies and museums to deepen our knowledge and appreciation of local traditions, and there are support groups and conferences for almost any particular group you can imagine. So when it comes to the care and feeding of particular stories there are now lots of new resources at hand. But there seems to be less offered in terms of new resources for exploring and affirming everybody's story. And this is because we are accustomed to looking to the resources of established wisdom traditions to help us in apprehending the universal story. The assumption has been that Christianity (or Judaism or Islam or whatever) *is* everybody's story. But this will no longer do. Whatever we once claimed for our particular religious traditions, and whatever dogmatists insist can be claimed for them now, we must finally admit that they are not qualified to tell everybody's story.

So where does this leave us? Who *is* qualified to tell such a story? Where do we go for a universal narrative account of how things are and which things matter? I believe there is no reasonable alternative but to begin with the sciences. The modern scientific account of how things are is our most valuable resource for telling everybody's story. Indeed, to a certain extent the narrative of modern science *is* everybody's story. A mere fifty years ago this assertion would have sounded ludicrous to scientists and nonscientists alike, for fifty years ago no one thought of science as a narrative enterprise. Science did not spin narrative accounts of anything—it just discovered facts, formulated laws, and constructed theories. If science had any value at all for the telling of everybody's story it was only because it could describe the arena, the scene, the stage on which the real drama of human life unfolds. Fifty years ago the sciences were engaged in the process of breaking down into smaller and smaller subdisciplines, each with its own separate domain to describe, and each with its own specialized vocabulary for describing it. But in the past fifty years there has been a truly astonishing reversal of that process. Instead of fragmenting into subdisciplines, the sciences are now merging into cross disciplines. The most exciting theoretical advances in science in recent decades are the ones that have managed to integrate the sciences of the large with the sciences of the small. In physics, for example, astronomy has been theoretically coupled with particle physics to produce quantum cosmology. In biology, evolutionary theory has been coupled with molecular biology to produce the grand synthesis. In psychology, neurology has been coupled with cognitive theory and other disciplines to produce the integrated cognitive sciences. Equally impressive are the theoretical links being forged between the major domains of science. Especially important are the links between biology and the physical sciences, on the one side, and between biology and the social sciences, on the other.

The result of all this theoretical work has been a gradual integration of the sciences. It is now possible, as never before in the history of science, to speak of a unified scientific worldview. Instead of generating diverse vocabularies the various sciences are becoming integrated by a common vocabulary.

What is this emerging scientific vocabulary? What has made possible the integration of diverse scientific disciplines? Just this: *the paradigm of evolution*. Fifty years ago the only science to speak seriously about evolution was biology. But today the paradigm of evolution is rapidly becoming the organizing principle for all the sciences—the physical sciences, the life sciences and the social sciences. Astronomers, physicists, chemists, biologists, psychologists, anthropologists—these researchers have come to recognize that they can no longer think constructively in their disciplines apart from the paradigm of evolution. The unifying insight behind this integration of the sciences is that the entire universe is evolving. The universe is a single reality—one long, sweeping spectacular process of interconnected events. The universe is not a place where evolution happens, *it is* the evolution happening. It is not a stage on which dramas unfold, *it is* the unfolding drama itself. If ever there was a candidate for a universal story, it must be this story of cosmic evolution. And it is only within the past two generations that the narrative features of cosmic evolution have come sufficiently into focus so that the story can be told.

Part I

HOW THINGS ARE

Prologue

In my youth I had an uncommon fascination with musical rounds. Part of the charm was the delight in discovering that a single phrase, layered upon itself, might have a harmonious result. Another element in the attraction was the intellectual satisfaction of keeping pace with increasing complexity. But perhaps the most captivating feature of all was the quality of self-containment. The round offered a world of sound sufficient unto itself. The cosmos is not a musical round, certainly, but there are aspects of cosmic evolution that evoke a similar fascination. The universe, like a musical round, is a self-contained entity of emergent splendor and complexity.

For most of its history the universe was void of life and consciousness. It was a monotony of matter. But then something entirely new emerged around four billion years ago when matter came to life. Living organisms manifest properties and principles that are not found in nonliving material entities. Yet living things are entirely material in the sense that differences between living things and nonliving things, as well as differences among living things, resolve into differences in the organization of matter. Life emerges like the second layer of a round—the quality of the sound is radically new, and yet it results from more of the same. And then suddenly (on a cosmic time scale) there appeared a third layer. Consciousness, like life, is a distinctive emergent reality—distinctive because it introduces unique properties and qualities, yet it is more of the same in the sense that all differences resolve into differences of material organization. Just as life is inconceivable apart from its material constituents, so consciousness is inconceivable apart from living matter. The universe *is* like a musical round—out of radiation emerges matter; through matter . . . life; through life . . . consciousness.

The perspective of cosmic evolution is essentially a materialist worldview, but it is a far cry from the old-fashioned materialism that simply declared that whatever is and whatever happens derives from the mechanical behavior of the generic material stuff of the universe. This classical form of materialism was, from its beginning, subjected to stern critiques from vitalists, idealists, and dualists. Vitalists argued that the attributes of matter could not give a coherent account of the phenomena of life, a failure that called for a sovereign life principle. Matter, they said, was categorically passive, implying that a material thing could be alive only if a transcendent vital force were imparted to it. Idealists and dualists made basically the same case for the phenomena of thinking, feeling, and willing. Inert matter was essentially incapable of producing these conscious events, which justified the assumption of mind (soul, spirit) as an independently real substance.

The perspective of cosmic evolution transcends the issues at stake between these classical worldviews, but does so without sacrificing the legitimate critical insights of vitalists, idealists, and dualists. In their zeal to safeguard the integrity of life and mind the critics of classical materialism argued for the sovereignty of nonmaterial substances. But the old materialism is long gone now, and there remains little need for these metaphysical safeguards against a new materialism that precludes the reduction of life and mind by its emphasis on emergent levels of organization.

Everybody's story is centered on a narration of the emergence of matter, life, and consciousness. In a nutshell the story goes as follows. As the universe evolved, matter emerged with properties enabling it to become organized in a staggering variety of complex patterns. The processes of life emerged out of these complex patterns. Living forms, too, came to be organized into various and complex species as the processes of life adapted to the conditions of changing environments. Among the many diverse forms of life are those endowed with a central nervous system, a structure from which emerged the realities of conscious experience, self-conscious behavior, and historical traditions.

What is so special about this story? What reason is there for thinking a story of such themes has potential for advancing global solidarity and cooperation? The very same reason there is for thinking *any* story might have this potential—that is, it has the power to engage the deep structures of human nature and to transform how we think and what we do. In other words, the narrative of cosmic evolution has the potential for harnessing the emotional effectors of kin selection and reciprocal altruism to serve the integrity of natural and social systems. Consider how. The story of cosmic evolution reveals to us the common origin, nature, and destiny shared by all human beings. It documents our essential kinship as no other story can do. This is no contrived shamanistic legend; this is not a bit of clever tribal tatooing—it is more like the real thing. This story shows us in the deepest possible sense that we are are all sisters and brothers—fashioned from the same stellar dust, energized by the same star, nourished by the same planet, endowed with the same genetic code, and threatened by the same evils. This story, more than any other, humbles us before the magnitude and complexity of creation. Like no other story it bewilders us with the improbability of our existence, astonishes us with the interdependence of all things, and makes us feel grateful for the lives we have. And not least of all, it inspires us to express our gratitude to the past by accepting a solemn and collective responsibility for the future.

It is possible, of course, to encounter the story of cosmic evolution without seeing any of this. There is nothing in the story to compell a desire for expanding solidarity and cooperation, any more than a slice of cake can compell one's desire to eat it. But the potential is there for those who will listen. Which brings me to a further point. I have made the rather bold claim that the story I am about to tell is *everybody's* story. Indeed, I believe it is, in the sense that it informs all human beings about our place in the scheme of natural history. But *my telling* of the story may not be to everybody's liking. It may be too superficial for some, too detailed for others; too florid for some, too arid for others. That can't be helped. What follows is one version of everybody's story, just as St. Matthew's gospel is one version of the Christian story. In order for the full potential of the story of cosmic evolution to be realized there must be many renditions of it—the more the better—at different levels of sophistication, with varying attention to details, and with divergent emphases. If the present version does not suit you then by all means set it aside and find one that does.

Now to conclude this prologue with nine simple queries.

1. *What is the universe?*
The universe is everything that is and happens—the totality of entities and events, including space, time, energy, and matter. To say that the universe is evolving is to say that it has undergone systematic developmental change through time, and to describe these changes is to tell the story of cosmic evolution.
2. *What are the gross features of the universe? That is, if we were able to view the universe from some vantage point beyond it, what would we observe?*
We would see an endless expanse, thinly populated by large lumbering structures, called galaxies, and even clusters of galaxies.
3. *What is a galaxy?*
A galaxy is an immense, loosely structured cloud of matter held together by gravity. The principal constituents of matter in these giant clouds are stars and interstellar dust. If we were able to go outside the universe we would see a dazzling array of these variously shaped star clouds scattered throughout an endless expanse of darkness.

4. *How large are galaxies?*
Very large indeed. There is much that is unknown about galaxies, but the most common estimates suggest that an average galaxy contains on the order of 100 billion stars.
5. *How many galaxies are there?*
Nobody really knows, but informed estimates run as high as 200 billion.
6. *What are these galaxies doing?*
They appear to be moving away from each other in a pattern suggesting they have a common origin. That is to say, the universe appears to be expanding outward from a single matrix. Astronomers have measured the rates of expansion and then worked their figures backward in time, as a way of calculating the duration of the expansion—which brings us to the next question.
7. *How old is the universe?*
The answer comes out to be about fifteen billion years. In other words, fifteen billion years ago the universe would have been very small. Leading theories suggest that at the time of creation the entire universe was compressed into a very hot and dense region, perhaps smaller than a single atom. The prevailing view is that the expanding universe we now observe is the result of a colossal explosion of this hot spot (the big bang), scattering matter and energy in all directions.
8. *What will be the final destiny of the universe?*
The final destiny of the universe is unknown. Everything depends on whether it is open or closed, and this depends on the total amount of matter in the universe. If the universe is open then it will continue to expand forever, in which case the final destiny will be an endless process of thinning and cooling. But if there is enough matter in the universe then eventually the force of gravity will slow the expansion until it grinds to a halt, whereupon the entire cosmos will begin to contract until it finally collapses into itself. That is, it will close up again into a superdense hot spot. If this happens then the possibility may exist for another big bang,

in which case the universe would expand again, then contract again, and so on—forever! Theorists refer to this possibility as the "oscillating" universe.

9. *Is the universe a self-generating entity, or is it the effect of some external causal influence?*

This is the God question, and I have every intention of dodging it for the moment. Whether or not there is an extra-universal causal influence is not an issue that can be resolved objectively, either by rational or empirical means. The question itself is both intelligible and enticing, yet at the moment the only way to resolve it is to draw on the subjective resources of one's own psychological state, or to submit to the authority of someone else's prejudice on the issue. Whether and how one produces an answer to this question tells us less about how things really are with respect to the universe than about how things are with respect to the subject. It is therefore a cosmological question having psychological (and perhaps social) significance. Perhaps an answer one way or another will matter profoundly to particular individuals. Perhaps, finally, a resolution to the question will matter profoundly to everybody. However that may be, it is clear that the issue belongs to a discussion of which things matter, and not to the question of how things are.

Chapter 1

The Organization of Matter

What Is Matter?

I once asked a physicist friend the simple question, "What is matter?" He responded with a glare of astonishment and then finally said, "You're kidding, right?" I was not kidding. I was quite certain that the question was intelligible and that any competent physicist would be able to give it a straightforward answer. I was mistaken. After my initial disappointment I made it a practice to put the question to every physicist I happened to meet. I have since had dozens of responses, none of them satisfying in the way I had hoped. *What is matter?* You cannot be serious. I do not have the slightest idea. Matter is what everything is. Matter just is. Matter is as matter does. Matter is a theoretical tool. Go ask a philosopher. And so the answers went.

It turns out that the nature of matter is not something physicists normally think about. They prefer to concern themselves with *states* of matter, not its essence. If you push hard enough a physicist may give you a circular answer. Matter is composed of subatomic particles that are themselves manifestations of energy—you know, $E = mc^2$. But at the end of the day it becomes clear that physics has no satisfying answer to our question. Physicists like to duck the question by insisting that it is a philosophical issue, not a scientific one. But philosophers are equally inclined to view it as a question for physics. One gets the impression that it is simply not the fashionable question these days.

From the seventeenth through the nineteenth centuries, however, the matter question was intensely fashionable as the mechanistic worldview of the new science was unfolding. At the heart of the mechanistic perspective was the corpuscular theory of matter, a slightly modified version of ancient Greek atomism. According to the corpuscular theory God had fashioned innumerable atoms at the moment of creation, each one solid, indestructible, and imperceptibly small. Atoms were thought to belong to various families, or species, each of these being present in nature by diverse and invariant proportion, and each characterized by unique geometric and chemical properties. That is, each species of atoms had its own essential nature, interacting with the forces of attraction to produce the various structures and qualities of the perceived world. While each atomic species (element) had its own peculiar properties, all particles of matter shared certain generic characteristics. They were extended in space, solid, massy, hard, impenetrable and movable. But more important, they were inert, passive, uncreative and soulless—entirely at the mercy of external "active principles" inherent in nature's laws.

The inertness of matter was a novel feature of the new mechanical worldview. Prior to the seventeenth century matter was thought to possess active properties, which meant that it was unnecessary to draw absolute distinctions of kind between inanimate, animate, and sentient beings. Distinctions between matter, life, and mind were not deep, they only *became* deep when matter was pronounced categorically inert. This new way of thinking about matter had the benefit of releasing scientific inquiry from the restrictive influence of theology, as the machine of nature could now be seen to crank along lawfully without need of continuous divine supervision. As liberating as the concept of inertness was for science, it meant that philosophy would be condemned to more than three centuries of toil over the mind-body problem. But that is another story.

Returning now to the point about the old-fashionedness of the matter question: What makes it old-fashioned? The question is old-fashioned because it appears to assume that there are some nifty little analogies on hand in terms of which we can form a satisfying answer. But there are no such analogies. The classical picture of matter was intelligible because matter could be pictured—atoms were tiny clods of hard stuff (somewhat like billiard balls). So we could understand. But contemporary physics does not lend itself to pool hall analogies, and all attempts to impose them succeed only in generating mind-numbing paradoxes. Quantum theory makes exquisite sense to the disciplined mathematical intuition of a few experts, but not to the rest of us.

The matter question has become increasingly elusive throughout the twentieth century. Already by the turn of the century the major theoretical moorings of classical mechanics had been abandoned, and by midcentury they had been replaced by an array of bewildering theoretical entities and dynamics. In the classical view matter was discrete and radiation was continuous, but according to the new physics matter can behave in wavelike patterns and radiation can behave like particles. In the classical view atoms were simple and solid units, but in the new physics they are complicated systems of interrelated events. A century ago atoms were the primary building blocks of material objects, but today more than sixty subatomic particles have been postulated. In the nineteenth century space was space and time was time, but in relativity theory space and time are inseparable. In classical physics matter was constituted in space, but in contemporary theory space is constituted by matter. Classical mechanics was committed to determinism, but quantum mechanics makes a principle of indeterminacy. In the old picture matter was matter, but in the new physics some matter is antimatter. In the nineteenth century matter was categorically inert, but today matter sizzles with bizarre agency. In the old days matter was full of theoretical integrity, but today, in the words of an eminent theorist, "matter is weird stuff." This is why physicists will stare at you in wonder if you ask them what matter is.

It would appear that Everybody's Story has got off to a disappointing start by asking the wrong question. But if we can't ask what matter *is* then what can we ask? We might fare better by shifting the focus of our inquiry from substance to process. All the indications are that we will understand matter best if we look at what it *does,* not what it *is.* Matter is what matter does. If we desire to know how things are then perhaps we should ask how they came to be.

How Did Matter Come to be Organized as It Is?

In the beginning was ultimate singularity. That is, before there was a universe everything that would become the universe was trapped inside a tiny morsel of incomprehensible heat and density. Before space or time, before matter and energy, before anything at all was the mysterious source: dark, quiet, profound unity. And then there was chaos, as the ultimate singularity released itself in the big bang—an event of such almighty force that it continues, even after fifteen billion years, to overpower the combined gravitational force of the entire universe.

Within a fraction of a microsecond after the moment of creation the forces of nature established their domains of influence and possibilities for future events began to fall within limits defined by the laws of physics. Space and time were themselves in the making, making way for the awesome events of unleashed energy. The universe expanded at an instantaneous rate, drawing out of the vacuum a sizzling gas of particles and antiparticles. In a blaze of spontaneous creation all forms of elementary particles split into existence together with their antiparticles, only to collide with each other and disappear instantly in the heavy traffic. A simultaneity of creation and annihilation produced an expanding chaotic fireball of blinding intensity. The paradox of this earliest phase of the universe is that the creation of matter was so prolific that most of it choked out of existence.

As it happened, there was a slight excess of particles over antiparticles, so as the fireball expanded and the collision rate dropped, there remained a sufficient number of unscathed particles for the universe to make something of itself.

Initially the universe was too hot for any physical structures to endure. But after about one second of expansion and cooling it became possible for free-floating quarks to join together to form neutrons and protons, which would later combine to form the nuclei of atoms. Still no atoms though. For several hundred thousand years the universe remained too hot for nuclei to succeed at capturing electrons. But when conditions were right copious amounts of hydrogen and helium atoms began to form, and for the next billion years or so the universe billowed forth in an expanding cloud of cooling gas. The organization of matter had commenced. The first phase of organization was extensive and enduring, but it was not complex. Hydrogen and helium are the simplest atomic structures we know. The heavier and more complex atoms were prevented their debut because the universe cooled too rapidly. Complex atoms require extreme heat for the necessary nuclear reactions to take place, but the window of opportunity had too quickly closed. Heavier atoms would have to wait for the heat of another day.

For the first billion years or so, the organization of matter was limited to the *microcosmic* assembly of simple atoms. But then the universe entered into a new phase of *macrocosmic* organization. The density of the expanding cloud of gas was not perfectly uniform. Some regions within the cloud were more dense and lumpy than others. Over time the influence of gravity took its effect by exaggerating the irregularities so that the cloud grew even lumpier. Different patterns of motion developed in distinct regions of the cloud, and slowly the gaseous universe began to fragment—in roughly the same way we observe rain clouds breaking apart in the sky. The result was that the original cloud distorted into distinct billows, and then gradually reorganized into separate clouds, each moving away from all the rest. These fragmented clouds—over 100 billion of them—were destined to become galaxies. The process of galaxy fragmentation was completed by the time the universe was a mere 5 billion years old.

Organization at the macrocosmic level continued. The initial patterns of motion that created the fragmentation process persisted in such a way that galaxies developed into many different shapes and sizes. And within each of these giant cloud galaxies there were pockets of greater and lesser density. In time, gravity exaggerated the irregularities until lumps of matter began to swell. As the relentless pull of gravity had its way, vast regions of atoms and elementary particles were drawn together with immense force. These regions became more and more dense with matter, and in the process the friction of colliding particles generated heat. So now we see a conspiracy of gravity, heat, and electromagnetic energy working together to give birth to the first stars. Stars are clouds of matter that become so dense and so hot that they ignite into nuclear furnaces. The whole process, from gassy lump to raging inferno, takes about ten million years.

There is nothing ordinary about a star, each one a dramatic contest between the inward force of gravity and the outward force of pressure. A star may hang in the balance for billions of years before one or the other force prevails. If there is sufficient matter contained in a star then eventually gravity will dominate and the star will collapse into a black hole. But if the pressures within prevail then the outer layers of the star will blow off in a giant explosion—a supernova. Meanwhile, during the contest of forces, the internal regions of a star can get hot enough for microcosmic organization to resume. When hydrogen atoms, helium atoms, and free particles become densely compacted under extreme heat the processes of nuclear fusion are triggered to synthesize the nuclei of heavier elements, from carbon all the way up to iron. Elements heavier than iron are synthesized in the explosion of a supernova. When a star goes supernova it splatters newly minted atoms into space to drift about as interstellar dust.

Microcosmic organization now continues apace. Free floating atoms of every kind are now available to obey the laws of chemistry, which means that under certain conditions they will combine with other atoms to produce a rich diversity of molecules. In recent years scientists have found evidence of highly complex molecules—hydrocarbons and amino acids—that have assembled themselves by chance out of the interstellar ashes of supernovae.

As a galaxy becomes strewn with the debris of supernovae explosions conditions begin to favor the construction of many new stars. Second- and third-generation stars form in the usual way—gravity collects matter into a ball until friction heats it to the nuclear flash point. Then it is a contest of the forces of contraction and expansion. This is precisely how our own sun was formed about five billion years ago. In the case of our sun there happened to be an abundance of interstellar dust in the vicinity that was prevented from collapsing into the fireball by a swirling motion. The result was a central ball of fire surrounded by a disk of orbiting matter. It is hard to say what caused this particular pattern of motion—some scientists hypothesize that shock waves from a nearby supernova could have done the trick. But whatever the causes were, the effect was to arrange the sun at the center with ten concentric swirling bands of matter at the outside. And it was from the matter in these swirling bands that the planets of our solar system were formed about 4.6 billion years ago.

Our solar system includes four rocky planets (the four bands closest to the sun), then a beltway of asteroids, and then four gaseous planets, and the icy Pluto. The rocky planets formed in roughly the same way that hailstones form, that is, by accretion, or aggregation. The earth, third removed from the sun, started out as a clump in the swirling disk, and as it swept through space its own gravitational field collected additional particles until it had vacuumed up most of the matter in its orbit. As the earth grew larger it became hotter (gravity, density, and friction again), eventually reaching the point where most of its matter melted down. This semifluid state allowed for a lot of shifting and sorting of the earth's materials. Heavy molten iron gravitated toward the center leaving lighter materials to be pushed toward the surface, resulting in concentric layers of iron and rock.

For its first 800 million years the earth remained very hot and was under constant bombardment by radiation and meteor showers (more bits of matter coming aboard). But then about 3.8 billion years ago the earth cooled, forming a rocky crust, called the *lithosphere*. Once the lithosphere became organized a variety of factors conspired to form the *hydrosphere* and the *atmosphere*. Radiation from the sun, condensation of water vapor, radioactive decay, and periodic outgassing from the still-molten interior produced an abundance of water and various atmospheric gases. Much of the earth's crusty surface was broken up and recycled in this process, leaving only a few fragments surrounded by water. The land masses we recognize today as continents are remnants of the earliest rocky crust. These land masses sit atop the active lithosphere, a system of eight to ten rocky plates that glide smoothly over the surface of the more plastic layers below. Over the eons the continents have ferried about the surface of the earth as the plates were mobilized by incessant motion within the fluid interior. Some 200 million years ago the land masses congregated in a supercontinent, Pangaea, only to separate again and take up their present—albeit temporary—positions.

The *biosphere* is the region of the earth's surface that supports life. More accurately, the biosphere is a highly complex geologically based biochemical system that developed out of the interactions between land (lithosphere), water (hydrosphere), and air (atmosphere). It was at the intersection of these major components of our young planet that microcosmic organization flourished to bring forth living creatures.

Interlude: The Grandeur and Grace of Matter

A few pages back I made mention of a "monotony of matter" to characterize the prelife period of the universe. Bad choice of words. Matter is not in the least monotonous. It is busy, creative, surprising, and melodic. If matter is as matter does, then matter is order-seeking, system-building, self-organizing, well-informed, excited stuff. And if modern physics had but one lesson to teach about this stuff it would be that matter is not to be underestimated, never to be taken for granted. Matter is just as grand as it can be. Still, nothing is more natural for us than to take matter for granted. After all, it does not seem very special. It is everywhere, it is everything. Matter is just as ordinary as it can be.

In the preface I said this book was all about leaving us with a sense of gratitude. And now I am at the point of suggesting it is appropriate that we feel grateful for matter. Gratitude, we saw, is an emotional response that evolved to regulate reciprocal behavior. We are moved to feel gratitude whenever we gain a favor or escape a loss, whereupon we find ourselves predisposed to reciprocate. I will address the question of reciprocity later on (including what or whom we have reason to be grateful *to*). But for the moment let us just consider what reasons there might be to feel thankful for matter.

Bear in mind that being grateful for matter cannot be meaningfully separated from a sense of gratitude for the entire domain of physical reality, that is, the whole universe. Matter, energy, space, time, and the natural laws governing these cannot be completely distinguished. Our question, then, is whether we have reason to be thankful for the universe.

What a silly question. Of *course* we do! Without the physical universe there would be no possibility of life, pure and simple. The lives we have are inconceivable apart from the physical universe that makes and sustains them. Nevertheless, even though we have sufficient *reason* to be thankful for the universe, we seldom have sufficient *cause*. Let me illustrate. Suppose you agree to do a favor for some business friends by hand delivering a large amount of money (say, $100,000) to their bank. You are going near the bank anyway, and there is no inconvenience involved. So you take the briefcase full of cash and off you go across town in your car. You finally arrive safely in the parking lot of the bank. Can you find reasons to be grateful that the trip has gone well? Sure, plenty of them. Yet you do not *feel* particularly grateful. That is, nothing in the reasons has caused an emotional response. But now suppose you step out of the car and there, teetering on the very edge of the roof, right where you carelessly left it, is the briefcase full of cash. Now you *feel* the gratitude.

The more we learn about the scientific account of creation the more we find ourselves left with such an experience. When you step out of your car to see the briefcase balanced on the edge of the roof your first thought is stunned disbelief. But there it is! Similarly, when we consider the odds against our universe producing the lives we have we feel more rational with the conclusion that none of it could have happened. But here it is!

For beginners let us go back to the creation of subatomic particles in the early universe. There in the big bang we saw that the universe expanded at an instantaneous rate, drawing out of the vacuum a sizzling gas of particles and antiparticles, simultaneously splitting into existence only to collide with each other and to vanish completely. We also observed that in this blaze of creation-annihilation there was "a slight excess of particles over antiparticles," leaving enough particles unscathed for the universe to make something of itself. It turns out that "slight excess" is more than a slight understatement. The excess was one in a billion. That is, for every billion antiparticles there were a billion and one particles. This means that if the early universe had been more evenly balanced by a factor of one-billionth then it would have been completely annihilated in the big bang. Odds like that make your trip to the bank look like a sure thing.

But if this is not enough, consider the expansion rate of the universe. The rate of expansion in our universe amounts to a sort of compromise between the explosive force of the big bang and the contractive force of gravity. On the one hand, if the explosive force were any greater (or if the gravitational force any less), then the universe would have expanded more rapidly than it did—too rapidly for any galaxies or stars to form. On the other hand, if the explosive force were any less (or gravity any greater), then the universe would have long since collapsed into a tiny cinder. So the universe we have (and thus the lives we have) is contingent on just the right balance between outward and inward forces. This is an astonishing fact, but it approaches the downright incredible when we consider the tolerance factor. That is, *how much* could the outward force or the inward force be varied and still result in a livable universe? The number works out to be one part in 10^{60}. In other words, by all odds we are not here.

If you are still unimpressed then consider the improbability of the periodic table of elements. We have the atomic elements that we do because the strong and weak nuclear forces are what they are. If the strong force were any weaker (by a minute fraction) then we would indeed have a monotony of matter, for the universe would be limited to hydrogen atoms. But if the strong force were any stronger then all the hydrogen in the early universe would have fused into helium—with the consequence that there would be no water, no stars, and no life. The fine tuning of the nuclear force is also relevant for the construction of heavier elements. If the strong force were any weaker then carbon atoms (the staple ingredient of life) could never have formed in the solar furnaces. But if it were slightly stronger then carbon atoms would have been fused into oxygen. There is no getting around it: the diversity of atomic elements is not to be taken for granted. Things could easily have been otherwise.

And now for the clincher. It appears that these highly improbable features of the universe are fundamentally independent, which actually multiplies the improbability of our being here. What are the chances for a coincidence of this *series* of improbabilities? Uncertain, but certainly very slight. It is as if a blind man, driving drunk through a war zone, were to arrive safely at the bank.

It strains the imagination and defies all rational expectation to suppose that improbability of this magnitude must be accepted as the brute fact of a chance universe. The difficulty of contending with the mental strain of such brute facts has encouraged speculative attempts to bring the odds into a more manageable range. One way of reducing the odds is to suppose the presence of transcendent purpose and design, that is, God. If the universe is seen as a mechanism for carrying out divine purposes then an unfathomable coincidence of chance events translates neatly into the intelligible consequence of creative design.

Another option for reducing the odds against our existence is to suppose a plurality of universes, the idea being that if our universe is one of zillions then extreme improbability resolves into eventual certainty. Wait long enough and *every* possible universe will have its turn. Several possibilities have been put forward to give imaginative substance to this option. One of these makes the conjecture that our universe is a momentary episode in a continuous oscillating series of expansion-contraction-expansion, and so on, *ad infinitum*. Another interesting possibility is that multiple universes inflate within an infinite expanse, each one with unique properties and duration, like so many bubbles in a boundless vat of oatmeal.

It appears, therefore, that we have one credulous option and two imaginative options for contending with the more puzzling aspects of the cosmic narrative. The credulous option accepts colossal improbabilities as brute facts. The theological-metaphysical option imagines a transcendent, or perhaps immanent, principle of intelligent design. And the pluralistic option imagines superordinate domains of space or time in which our universe appears as one among many.

Each of these options is burdened with immense intellectual difficulties. And while there are at present no objectively compelling reasons to prefer any of them, one can easily see how various subjective factors might draw individuals to each. What is more difficult to understand, however, is how anyone could suppose that the deep mystery behind the cosmic narrative might be diminished by the imagination. In this respect both theology and speculative science are deluding themselves by pretending to take us beyond the strain of the credulous option. A brute fact is a brute fact, and as brute facts go transcendent deities and multiple worlds are inherently no easier to swallow than impossible odds.

At the end of the day there remains the grandeur and the grace of the universe, full of enduring promise, inviting us to relent in wonder at the mystery of it all. Here is a mystery to command our curious efforts to understand, but here too is a mystery to provoke in us the raw sense that we are blessed.

Chapter 2

The Organization of Life

What Is Life?

If the life question is to be properly understood it must be approached as a variation on the matter question, for life is simply one more thing that matter does, given the right circumstances. The processes of life bring into being new properties and phenomena, for sure, but these are nevertheless material. Life is an emergent reality, not a transcendent one.

Unlike the matter question it is possible to find a satisfying answer to the life question. Such an answer, it seems to me, would have to be a compound answer—one that tries to integrate the material aspects of life with its functional aspects. That is, to describe what life is we must try to say what it is made of and what it is made for. Once these attributes of life are clarified we will be well prepared to ask how life came to be organized as it is.

A complete material definition of life would soon exhaust us with detail. One might begin with a simple list of the eighteen elements that have become involved in the chemical composition of living organisms. Of these eighteen elements, four of them (hydrogen, oxygen, nitrogen, carbon) make up about 95 percent of all biomass. From a list of basic elements our material definition might begin to branch off to describe various groups of essential compounds, such as amino acids, nucleic acids, proteins, and so on, until the definition included a very large number of specific ingredients. To be fully complete the definition would then have to catalog the materials essential to each particular species of life, and then to describe the intricate anatomical structures into which these materials are integrated. In the strictly material sense, then, life *just is* basic atomic elements organized into molecules, then into macromolecules, then into organelles, and then cells, tissues, organs, and so on. The point being that living organisms, even the simplest ones, are material constructions existing in highly ordered states.

A complete material definition might bring us a long way toward understanding what life is, but a satisfying answer to our question would have to include the principal functions of life as well. It is not quite enough to know that living things are complex biochemical systems—we also want to know what they are doing. So our next step is to inquire about those functions without which life cannot exist. At the level of particulars, of course, there are lots of these essential functions (which means there are lots of ways to die). But we are here concerned with the essential functions of life at the most general level.

Perhaps some insight may be gained by asking first about death. What happens when an organism dies? The obvious thing to say is that when an organism dies it begins to decompose. At the most basic level, then, the principal function of life appears to be that of keeping the rate of chemical composition well ahead of the rate of chemical decomposition. An organism failing to do this is either dead or soon to be so. Yet too much chemical composition cannot be a good thing either. Chemical composition has to be very precisely regulated—too little and the organism will decompose, too much and it may begin to consume itself. Life depends on precise and sophisticated chemistry. Consider: within the membrane of a one-celled creature there occur on the order of a few thousand different *types* of chemical reactions. And most of these reactions are continuous, so that in the span of a few seconds millions of chemical reactions take place. Imagine what a detailed account of chemical reactions in a multitrillion celled human body would be like!

A rough answer to the life question says that living things are self-regulating biochemical systems that manage somehow to keep themselves organized against the odds presented by a harsh and pitiless environment. But this tells us little. We still do not know the basic means by which they do this. How do biochemical systems keep the rate of chemical composition comfortably ahead of the rate of decomposition? The answer is that they do it by transforming energy and by processing information.

Every living thing is both an energy transformer and an information processor. These two general operations work in tandem to build and maintain organisms. Information processing consumes a lot of energy, but with the right kind of information at hand an organism can get all the energy it needs by using the information to construct chemical machines for trapping energy. So if an organism has the right combination of information and energy then it can continue to metabolize. It would appear to follow that any chemical system that could *start out* with a supply of energy and information would be well equipped to keep itself going. It is the exploitation of this principle of inheritance that accounts for the origin and diversification of life. Life as we know it emerged from nonliving matter when chemical systems integrated the functions of energy transformation, information processing, and self-replication.

How Did Life Become Organized as It Is?

When the announcement finally comes that life has been artificially brewed from nonliving matter very few biologists will be surprised. They will be intensely curious about the actual mechanics of the event, but unmoved by the fact of it. It is generally assumed in the scientific community that living forms emerged spontaneously from nonliving organic molecules, which in turn assembled from inorganic matter. The potential for life is inherent in the properties of matter. It is all a question of organization.

Until the details are fully known we shall have to be content with a "likely story" about the origins of life. Various scenarios have been put forward, a leading candidate being the "RNA world" hypothesis, which postulates the emergence of life from a prebiotic soup about 3.8 billion years ago. The prebiotic soup was a chemical quagmire of molecular evolution wherein molecules competed with one another for the attention of atoms. Nothing very mysterious here—it just so happens that some molecules are more stable than others, so as fragile molecules broke down their atoms would be free to join in the construction of more stable molecular units. The big winners in this process would be selected for having special properties, such as those for catalyzing chemical reactions (metabolism), making copies of themselves (replication), or both.

These were the conditions favoring an RNA world. Primordial RNA molecules were the biggest winners in the selective process because, as the story goes, they were capable of both operations. In their folded state RNA molecules would be effective catalysts, and in their elongated state they would code for self-replication. So when folded RNA acted on elongated RNA the result was a world of increasing abundance of RNA supermolecules. With the rudiments of metabolism and heredity already mastered by the RNA world it is not difficult to imagine further elaborations of these functions. Once protein synthesis and the DNA molecule enter the picture all the dynamics would be in place for the graduation of some chemical systems to the status of biological systems.

The RNA world is not the only candidate for describing the emergence of living systems from chemical systems. A rival hypothesis contends that the phenomena of metabolism and replication emerged independently in the prebiotic soup, eventually coupling to form living cells as a result of chemical parasitism. The picture offered here is that proto cells, capable of metabolism and protein synthesis, were invaded by molecular parasites proficient at self-replication. In time the parasitic relationship stabilized to a symbiotic relationship. Within the protective environment of the cell membrane chemical fine-tuning proceeded apace to construct the genetic mechanisms (RNA, then DNA) that allowed metabolizing cells to grow and divide with great precision and regularity.

We may never know the full story of how living systems emerged from nonliving material. What we do know is that the event happened within a billion years of the earth's formation. Once the functions of metabolism and heredity were coordinated within the living cell conditions were set for unfolding the long series of experiments in biotechnology we know as biotic evolution. We owe the rich diversity of life forms on earth to a pair of negative facts: first, the genetic machinery locked in the nuclei of cells is not invulnerable to error, and second, the environment into which the first cells emerged was not homogeneous. The first of these facts accounts for *genetic variation* and the second accounts for *natural selection*. When copy errors (mutations) occur in the process of replicating the genetic code new instructions are carried forth resulting in novelty in the traits of organisms. Any new trait, structural or behavioral, may be either beneficial or deleterious to an organism as it negotiates a living among the features of its natural environment. Random variations in genetically endowed traits are thus constantly being interfaced with a changing environment. Given enough of this chancy sort of thing, the inevitable happens: there will emerge lots of different ways to be a living thing.

The first living cells on the planet basked in an economic utopia, one in which environmental conditions allowed everyone to consume at will while no one was constrained to produce. These single-celled organisms made their living by fermenting energy-rich nutrients that had been cooked up in the primordial soup. At first there was more than enough for everyone, but eventually the increasing population of consumers brought the primordial community of life to the brink of ecological disaster. Under these increasingly competitive circumstances mutations resulting in new traits for exploiting alternative energy sources would have been favored. Some organisms inherited strategies for devouring the remains of dying cells, thereby constructing the first links in a biotic food chain. But these carrion strategies could do little more than buy time in the face of an absolute decline in the primary sources of food.

The real breakthrough came with a series of mutations leading to photosynthesis. Photosynthesizing organisms are able to utilize energy from the sun to manufacture their own food supply, thus freeing them from dependence on ready-made energy locked in the dwindling supplies of molecular food. But photosynthesizers offered a literal opportunity of a lifetime to nonphotosynthesizers as well—that is, they stored energy in the form of sugars, which would nourish organisms capable of eating photosynthetic bacteria. The advent of photosynthesis established an organic infrastructure for a dazzling future in biotechnology. Stimulated by the assaults of ultraviolet radiation (no ozone shield at the time), mutant organisms fashioned new metabolic pathways to form interactive food chains throughout the marine environment.

There was, however, a downside to the invention of photosynthesis. The principal byproduct of this chemical reaction is free oxygen, a highly toxic substance to most species living in the primordial environment. It was becoming apparent that the process of evolution had solved the problem of food shortage only to create a problem of pollution. Life had turned into an exercise in self-destruction: the more photosynthesis, the more poisonous oxygen. The whole experiment of life might have choked itself out had it not been for mutations resulting in the process of respiration. Respirating organisms are oxygen tolerant, but far beyond a mere tolerance of the stuff they are capable of exploiting its explosive properties for the purpose of more efficient metabolism. A substance that once threatened life with its potential for accelerating chemical decomposition now came to enhance life with its potential for accelerating chemical composition.

The abundance of free oxygen in the earth's atmosphere not only created the conditions for new and highly efficient metabolic pathways, it also prepared the way for living organisms to explore new geographic domains. Enter (ever so slowly) the ozone shield. Prior to the development of the ozone layer the earth was bombarded with dangerous ultraviolet radiation. This meant that living beings were constrained to life at sea, where they were protected by the light-filtering properties of water. But as free oxygen drifted skyward it was reconfigured by solar energy into ozone (O_2 + sunlight —> O_3). Ozone is a life-threatening substance on earth, but in the upper atmosphere it protects life by shielding out lethal ultraviolet radiation.

The ozone shield did not assemble itself overnight—it took eons worth of photosynthesis to produce enough oxygen to blanket the earth. Thus the span of time between the invention of photosynthesis and the emergence of life from the sea was immense. The first stirrings of life began 3.8 billion years ago, and photosynthesis started up shortly thereafter (about 3.5 billion years ago). But it was not until 600 million years ago that living forms ventured forth upon the land. A full five-sixths of the earth's biohistory has been lived out exclusively in the oceans.

The biotic colonization of land surfaces takes us well ahead of our story. A lot of complicated organization had to take place in aquatic environments before this happy event could occur. So while the ozone layer is slowly forming to make the land safe for life, we might return to the sea for a review of three major breakthroughs in the organization of life: the elaboration of cell structure, the invention of sex, and the emergence of multicellularity.

Two billion years ago there was a single form of life: bacteria. The architecture of these cells was quite simple—each bacterium had a cell wall containing the unbounded DNA molecule and a supply of chemicals necessary for metabolism. But then, perhaps 1.5 billion years ago, there emerged a fundamental divergence between two cell types: the prokaryotes and the eukaryotes. Prokaryotes descended from the ancestral bacterial cells, noted for pioneering the basic forms of nutrition and inventing a broad diversity of metabolic pathways. The newcomers, the eukaryotes, specialized in morphological diversity. Eukaryotes distinguished themselves by having discrete internal structures (organelles), membrane-bound components for specialized cell functions. These larger and more complex eukaryotic cells may have evolved as a result of symbiotic mergers whereby certain bacterial cells found ways to take up specialized residence within superordinate host cells. But however this crucial "cell divide" came about, the important breakthrough is that eukaryotic cells developed the ability to arrange their internal structures in various ways, resulting in a greatly increased potential for diversifying their shapes, sizes, and movements—all variables that prepared these cells to adapt to new environmental niches.

The most common pattern of eukaryotic cell reproduction is described by the process of *mitosis*. In mitosis the DNA replicates itself within the cell nucleus, after which the nucleus divides into two identical nuclei, whereupon the cell pinches at the middle to produce two daughter cells, each an exact copy of the parent cell and each endowed with potential for producing the next generation. A major breakthrough in biological organization came with the invention of sexual reproduction. The details of this event are lost to the obscurity of natural history, but at some point, about one billion years ago, some eukaryotic cells found a way to produce slightly varied and complementary offspring. When these complementary cells met by chance in their fluid environment they fused together to make a single cell having a mixture of DNA from each "parent" cell. In such cells there followed a pattern of division known as *meiosis*, which sorts out novel combinations of genes into four daughter cells, each of which can reproduce only by merging with a complement cell in the event of fertilization.

Sexual reproduction accelerated the diversification of life forms and made possible true multicellular organisms. The multicelled organization of life became possible when some organisms developed the capacity for orchestrating both meiotic and mitotic patterns of cell division. In these organisms we observe a fundamental distinction between cell lines. Along one line (the *soma* line) cells multiply to form body tissues by simple mitosis, and along the other line (the *germ* line) cells reproduce by meiosis and fertilization. Descendants of these organisms could then grow bodies of various sizes and shapes by the process of mitosis, but they could also reproduce genetically novel (thus diverse) offspring by the sexual strategy of meiosis-fertilization. These sexually reproducing organisms had to be able to negotiate their way through two major challenges. First, they had to possess traits sufficient for exploiting the immediate environment for chemical resources, and second, they had to possess traits for successful mating. Individual organisms inheriting the best combinations of morphological and behavioral traits would succeed at both tasks, and thus genes for these traits would be recombined by the mechanics of sex, then to be wagered against the challenges facing the next generation.

Multicellularity caught on because it conferred enormous advantages in terms of predation and motility. The first multi-celled organisms were probably aggregates of virtually identical cells. In time, however, soma lines diversified to construct specialized tissues and organs that proved beneficial in negotiating new environmental niches. And by 600 million years ago, when the ozone shield was securely in place, the potential for new niches had increased dramatically. Now conditions were set for a riot of biological organization to take place. New species sallied forth upon the land, both to occupy and to create new ecological niches. The lesson of the past 600 million years has been that there are lots of ways to be a living thing. Since the first stirrings of life on the planet, the evolutionary process has organized well over a billion species. Most of these life forms (99%) have gone extinct, but there remain today somewhere in the range of five to thirty million species.

There is much more to the organization of life, however, than one finds in living organisms. Ecosystems and bioregions—and, indeed, the biosphere itself—are not living beings, yet they are highly organized natural systems that result from, but also enable, the flourishing of life. These systems are not life forms, exactly, but they are clearly *life-formed*. They are emergent realities, whole systems with dynamics of their own, not reducible to their constituent parts. Again, we see a familiar theme: the integration of constituent matter into coherent systems. A multicelled creature is an emergent unity of systematically organized component parts, that is, divergent cell lines. But the component cells are themselves unified systems of discrete organelles. And each organelle is in turn a coherent arrangement of interacting molecules, which are systems of atoms, which are systems of subatomic particles. And then there are ecosystems, bioregions, and the biosphere—each of these described by systematic interactions between living things and their nonliving environment. And all of this organization unfolds within the macrocosmic systems of planets, stars, and galaxies. There are systems all the way down and all the way up. Emergent systems properties feed back on existing systems, changing patterns of interaction until new components arise to be captured by order. This is how things are.

Interlude: The Wonder and Wisdom of Life

A mutation is a small thing. By the standards of speech a mutation amounts to a simple mistake, a mere slip of the tongue, or an occasional typo in a printed book. Mutations are copy errors occurring in the replication of DNA. The important point is that even trivial copy errors can have dramatic effects on the chain of cellular command, from DNA to RNA to protein synthesis. So when a mutation occurs it typically shows up in terms of protein function, which ultimately affects the properties of an organism's cells.

Mutations may be small things, but they can have immense consequences. For example, a single strategically located genetic mishap can be lethal to an organism. A slight change in the chemical syntax of an oncogene somewhere in the body might trigger the production of a hyperactive cancer-causing protein. A mere slip of the genetic tongue could bring down a majestic elephant. Small things, as it happens, can command a lot of respect.

The most respectable mutations—that is, the ones having the greatest consequences—occur in the germ lines of sexually reproducing organisms. Here the results of copy errors are magnified by their potential for being passed on to offspring. A mutation in the sexual materials of one generation may show up in the way certain proteins function in the next. Such changes might produce new traits that are then tested against the features of the environment. In the ongoing process of evolution beneficial genes will thus be selected into the gene pool of a species while detrimental genes will be selected out. Here we catch a glimpse of the awesome power of genetic mutation and natural selection to craft new forms of life. Mutations embellish and the environment chastens. Over time, the litany of genetic addition and environmental subtraction has produced a wonderment of diversity and complexity of living systems. Small things indeed!

When one considers the astronomical number of small things that had to accumulate for us to have the lives we do, it is tempting—some would say compelling—to suppose that the whole epic of evolution was deliberately and ingeniously planned out to execute an ulterior divine motive. This is a perfectly natural supposition. After all, our brains are specialized for discerning the motives and plans lurking behind the actions of other beings, especially those of our own species. We are, it appears, experts at mind reading. What could be more useful in the ancestral environment than the ability to predict how another agent would behave? And what more useful for this prediction than a well-informed hunch about that agent's motives? Behaviors become intelligible to us in the measure that we can construct psychological scenarios for them, that is, when we come to see them as motivated acts. Therefore, when such a reasoning device as the human brain is brought to bear on the wonders of the living world we should not be surprised if it concludes that somehow lurking behind the curtain of nature there is an awesome will and wisdom.

There are, of course, many who oppose this conclusion by insisting that it is both fallacious and unnecessary to map psychological scenarios onto the phenomena of nature. Whatever evidence might be taken in support of divine agency can be explained sufficiently without it. But either way—whether one ultimately explains the organization of life by the willful designs of intelligent agency or by the blind chance and necessity of natural mechanisms—the brute fact remains that there *is* organization, there *is* a logic. Whether within nature or behind it, there is an irresistible wonder and wisdom to life.

It is interesting to note that both sides in the debate employ the principle of reverse engineering in the pursuit of explanation. Reverse engineering is a method of reasoning backward from the observation of an orderly system to some principle of design that might explain it. Thus if you are an archaeologist and you find some sort of prehistoric tool, you might use this backward reasoning to conjecture about its function. Those motivated by a theological agenda will reason back from biological systems to God's plan, while those having naturalist assumptions will reason back to the algorithms of nature. Richard Dawkins couples this exercise in reverse engineering to the economic principle of *utility function,* a term referring to "that which is maximized." These ideas in tandem encourage the investigator to probe nature by asking what properties living systems are designed to maximize. The answers one comes up with should reveal something profound about the wisdom of life.

The place to begin this inquiry is not with living systems but with the universe as a whole. What is the universe designed to maximize? The answer is given by the second law of thermodynamics—the so-called law of entropy—the most inexorable of all physical laws. "Entropy" means, literally, a transformation of energy from a workable state to a nonworkable state. This is the law that prevents the possibility of any perpetual motion machines. The discouraging thing about these machines is that no matter how well they are engineered and lubricated they always generate friction, which amounts to energy escaping the system in the form of heat loss. Progressively, then, the system has less and less energy available to do work until eventually the motion ceases. Entropy is a measure of the energy lost in the process. In truth, the energy is not lost—energy *loss* is prevented by the first law of thermodynamics—it is just rendered unavailable or useless to do work. Entropy—that is, energy rendered useless—therefore increases in such systems over time. The second law applies to the universe as a whole, which means that the total amount of useful energy in the universe will decrease over time, or in other words, entropy will increase. A reverse engineering analysis might well conclude that the universe is designed to maximize entropy. This would be the irony of ironies, for it implies that the utility function of the universe is to become completely useless.

But what about the utility function of life? The course of biological evolution appears to indicate that living systems violate the law of entropy with impunity. While inanimate systems slide obediently downhill toward a state of disorder, life appears defiantly to reverse the process, sending energy uphill toward more complex ordered states. If the universe maximizes disorder, life appears to maximize order. Creationists have seized on this apparent contradiction to infer divine intervention on behalf of the life process. Since life is central to the divine plan, God has seen fit to muzzle the jaws of entropy for as long as it takes the drama of salvation history to run its course. The standard reply to this theological inference is to point out that the law of entropy does not directly apply to living systems since these are "open" subsystems, continuously drawing new energy from the outside (i.e., from the sun). But since the universe, a closed system, does not have an external energy source the second law holds.

A deeper account would be to recognize that living systems are themselves highly efficient mechanisms for transforming energy into a useless state. In other words, life does not momentarily and paradoxically retard the increase of entropy in the universe, but actually *accelerates* it. Here is yet another deep irony adding to the mystery of the universe: the fast track to disorder is through a concentration of order. On this view the organization of life may be seen as a function of the universe to find ever-more creative ways to comply with the iron law of entropy. Indeed, creativity and entropy are closely related by a theoretical association of entropy and information. The creativity of life is brokered by information at the expense of useful energy. Alternatively, energy is rendered useless by its informed use in living systems.

The utility function of life, then, is to accelerate entropy by extracting information from energy, that is, by maximizing the amount of DNA. The consequence of this utility function has been the organization of life into a complex network of hierarchical systems. DNA is information for orchestrating the construction and behavior of biochemical machines. These machines are designed to exploit their environments for the resources essential to producing more DNA. Living forms are themselves important factors in the environments to be negotiated by other DNA-bearing machines. The logic of life takes this into account, investing in more and more complex machines capable of exploiting the hard-earned chemical achievements of simpler ones. This explains your being spared the inconvenience of photosynthesizing—your machinery shortcuts the process by exploiting the *products* of photosynthesis. It also explains why your body does not have to bother with the construction of various complicated organic compounds, like vitamins, which are precooked for you by the life forms you eat.

We have the wonderful lives we do because of thousands of life lines to other organisms at all levels in the organization of the life community. Each one of us is wired into an immense web of interconnected living parts, where each part is an award-winning systematic design—and the whole thing has been jerry-rigged over billions of years by the accidental wisdom of DNA. Once again, it is tempting to use metaphors of intelligence when trying to apprehend even the possibility of something as complex as the community of life. But there are dangers inherent in such metaphors. After all, it is the evolutionary process that produced intelligence along the way, as one more strategy for maximizing DNA. To read intelligent design back into the intelligence-designing process is to attribute the wind to the windmill.

No, there is something far more subtle and perplexing going on here—some elusive pattern-seeking that mysteriously draws something more out of nothing but. The process is too ridiculously stupid to be called intelligent. What measure of intelligence would endorse a process of mutation-selection that gets mostly way-wrong answers on the design problems? And yet when the few workable solutions do appear they are completely unpredictable and transcendently ingenious. There is a patient prospecting wisdom at work in the emergent organization of life. From a hillside of random copy errors a few small nuggets of information are panned out for safekeeping in the genetic code. And by the accumulation of these small things an exquisite lattice of life has come forth to embrace the earth.

That such an ad hoc and haphazard process has created the most complex and intricate designs in the universe is the most stunning and ingratiating fact one can behold. Quite apart from the meaning one attaches to it, this fact alone compels the humility and gratitude of every person alive. By an accidental wisdom we are granted the lives we have.

The mystery deepens as we come to realize that our own death is implicit in the process of life. It is quite natural to look on death as a problem. Indeed, in most cultural traditions death is *the* problem of human existence. But from the perspective of evolutionary wisdom death is not a problem at all, it is a *solution*. And a fairly recent one at that. Prior to the invention of sexual reproduction there was nothing inevitable about death. It happened all the time, of course, but it did not need to. Many species of organisms need never to die. Single-cell organisms reproduce by simply dividing into halves, each half becoming a distinct individual capable of further subdivision. Death is not a part of the picture, for both halves go on living, enjoying a virtual immortality. Even in multicelled flatworms the prospect of growing old and dying makes no sense. They merely pinch at the middle, leaving two abbreviated flatworms to regenerate all the missing bits—and so on, indefinitely. Death has no sting to a flatworm.

It is only among the more complex sexually reproducing organisms that death enters the picture as a certainty. The reason has to do with the divergence of cells into the germ line and soma lines. The germ line produces informed seed for the next generation, while the soma lines diverge for the production of various body parts. The body negotiates the environment, thus enabling the germ line to do its own job of bringing forth more seed bearing organisms. The strategy is simple and elegant: the soma line is the instrument of the germ line. It is all part of the utility function of maximizing DNA. Having performed its duty to the germ line, the body becomes redundant and eventually dies. But the germ line continues immortally onward in subsequent generations. The death of the body is an essential part of the design.

In the wisdom of this scheme it becomes difficult to view death in a negative sense. The inevitability of my death is now beheld as a necessary condition of the life I have. A mere entrance fee, to be paid on the way out. If there were no death there would be no soma line, and without a soma line there would be no possibility of an embodied person—no memories, no loves, no joys, no wonder or wisdom, no longing or learning. These are among the splendors of the body, and for these we must die. We *must* die because we *get* to live. To the extent that I cherish my life, therefore, I have reason to be profoundly grateful for my death.

How then shall I think about death? With gratitude, and as the occasions decide. When I have occasion to mourn the death of others I will try to absorb the loss in what I have gained from them. I will try to understand my grief as a measure of my gratitude. And when I have occasion to consider the fact of my own death I will attempt to think large. I will try to see that a soma-centered story of the self is a small and impoverished view, and that the life within me was first quickened among the primordial organisms appearing on earth four billion years ago. I will affirm that all lives, no less my own, are instruments of life itself. And by these measures I will submerge the absurdity of death in gratitude for the wonder and wisdom of life.

Chapter 3

The Organization of Consciousness

What Is Consciousness?

Our story opened with the big bang and proceeded to narrate the organization of galactic, stellar, and planetary systems in accordance with the laws of physics and chemistry. Then we saw how these laws allowed for the emergence of new properties and principles of material organization in the form of biological systems. And now we are about to survey the emergence of consciousness from the preconscious dynamics of living systems.

If we are not careful in our thinking about consciousness we might very quickly find ourselves lost in a muddle. And once muddled about the nature of consciousness we may, as so many others have, simply throw up our hands and declare it to be the mystery of mysteries, a reality so far beyond our comprehension that it defies all attempts at definition. What is consciousness *really*? Is it a flame? A dance? A stream? Is it the ground of all reality, as the idealists believe? Is it a place (a sensorium) where events occur? Is it an irreducible substance to be categorically distinguished from matter? Does it possess powers of causation? Is it subject to laws of nature? Or do such questions merely invite confusion?

The best way to avoid certain muddles is to admit that there exist a good many questions that are simply undeserving of answers. Our question about the nature of consciousness ranks first among them. There can be no satisfying answer to the question. But this is not because consciousness is the deepest of all mysteries. Just the opposite. It is because there is absolutely nothing mysterious about consciousness to begin with. Ask a neurologist about the nature of pain and you will get a definition that bears no useful relation to your experience of pain. But of course the mistake is to ask for a definition in the first place. Anyone who has experienced pain will have no need for a definition of it. Likewise with consciousness. If you are conscious you already know everything there is to know about the ultimate nature of consciousness. Definitions are pointless and unsatisfying. Anyone at all who might possibly benefit from a definition of consciousness does not need one—the reality itself is there, immediate and complete.

The interesting (and deserving) question about consciousness is not what it *is*, but rather how it comes to be. When we ask about the nature of consciousness what we really want to know about are the conditions under which it occurs. The details of these conditions are still largely obscure, but there is good reason to feel confident about the general picture.

Conscious experience is produced by the organized activities of a central nervous system, by the brain in particular. No brain, no consciousness. The brain functions to process neural signals. It receives input signals from upstream sensory neurons, alters and enhances them by the activities of interneurons, and then transmits output signals to downstream motor neurons. Somewhere in the process of interneuronal enhancement conscious experiences are generated. But how?

It is useful to think about brain function by analogy to the logic of Darwinian evolution, that is, variation winnowed by selection. Consider the activity of brainstorming. In brainstorming sessions we deliberately generate a range of options for behavior, the more numerous and far-fetched the better. This stage parallels the generation of genetic diversity in the evolutionary process. But then the brainstorming project turns to the task of evaluating options with a mind to ruling out the bad ideas, an operation analogous to environmental selection. At the end of the selection process we are left with the survival-of-the-fittest option for behavior. In this view the brain functions both to generate and to limit diversity among options for behavior.

The brain routinely operates as a high-speed brainstorming device. As we interact with our environment we are under continuous demand to monitor the world and to design behaviors appropriate to it. Information from sensory systems, association systems and long- and short-term memory systems construct for us a kaleidoscopic stream of images about how things are out there in the world. There follows a random generation of imagined scenarios for what might happen next. Fragments of information from attention and memory are associated instantaneously to propose a broad range of guesses about how things are going. All of this activity transpires well beneath the level of conscious experience. The preconscious brain is an unpredictable storm of unresolved possibilities, both for what the organism will perceive and for how it will react. The orderliness of conscious experience emerges out of the brain's activity in generating the storm to begin with, and then inhibiting some possibilities in favor of others. There may be thousands of possibilities vying for conscious attention at any given moment, each of them being scanned and graded by emotional markers stored in memory and reflex systems. Whatever is left surviving at the end of the editing process becomes the featured stuff of consciousness.

The general picture, then, says that conscious experience is what results when a rich hierarchy of neural systems transmits signals at a very high level of complexity. It would be quite another matter to specify just what that level is. Perhaps there is some magical threshold of signal enhancement that must be crossed before the rheostat of conscious experience clicks in. If this is the case then we might wonder which animal species are capable of operating above the threshold. Mammals and birds, for sure. But what about fish, reptiles, and insects? Are these organisms perpetually preconscious because they lack the neural subsystems required for an array of options for perception and behavior? Can we assume that they are periodically sentient but never continuously so, as conscious beings are? And what about machines? If we understand consciousness to be a function of information processing then what grounds in principle do we have for rejecting the possibility that computers might one day transcend the magical threshold? The search is on for satisfying ways to think about these issues. But for the moment, at least, it seems right to say that conscious experience is contingent on a certain amount of plasticity in central nervous systems—plastic systems being those having high potential for creating and assessing possibilities for alternative states. The evolution of this plasticity amounted to the emergence of radically new properties of matter together with new principles and dynamics of organization.

How Did Consciousness Become Organized as It Is?

Multicellularity is a terrific door opener. If you have an abundance of cells then you can dispatch diverse cell lines to carry out specialized functions, thus to negotiate forbidden niches and gain access to new sources of energy. But there is a price to pay for multicellularity: it requires something like government regulation. That is, in order to get many cells to behave coherently, as a single entity, there must be mechanisms for the coordination and control of subsystems. Complex multicellular organisms have developed specialized cell lines for these regulating functions.

The largest organisms on the planet are trees, some of which have carried the multicellular idea to extremes. A single Quaking Aspen, for example, may send offshoots forth to command a territory of several acres, growing to a biomass equivalent to dozens of blue whales. Trees, like other plants, handle the regulatory problems of multicellularity by developing vascular systems to carry chemical information to various parts of the organism. A vascular system will do nicely if you have the patience of a tree. But most animals need information mechanisms that are faster and more versatile. For these purposes the neuron is just the thing.

Neurons are the liveliest of cells. Like other cells they perform the usual functions of protein synthesis and metabolism, but neurons are uniquely capable of becoming excited. And embedded within the patterns of excitement is information essential to the normal functioning of the organism. In addition to their excitability neurons are uniquely well structured for multiple interactions with other cells. The typical neuron, for example, will be capable of interacting with up to a thousand cells. Put 100 billion neurons (the human count) into hundreds of interactions each and you have the potential for trillions of information events. The organization of such complexity is not a simple matter.

Much of the organization of neural complexity has been achieved by the accidental wisdom of natural selection. As multicellular animals occupied terrestrial niches, and as new vectors of competition and cooperation developed, there were selective pressures favoring more sophisticated sensory and motor systems. Arms races between predators and prey would reward sensory designs for detecting the whereabouts of other organisms. Any breakthroughs in processing information about motion, distance, color, sound, smell, and so on, would have survival value and would become fixed in the genomes of various species. Adaptations on the sensory side would tend to be evenly matched by new motor traits as well. New sensory powers would have limited value if the information could not be used to make a catch or to avoid one. So the sensory arms races would be balanced by motor arms races favoring greater speed, accuracy, strength, agility, and so on.

But certainly neither of these branches of arms races could go forward without the logistical supports of increasingly sophisticated neural systems. Input patterns would have to be assembled and evaluated before output patterns could be constructed and released. Particularly important was the evolution of complex interneuronal systems capable of integrating different forms of sensory information into coherent perceptions. Animals capable of associating visual and auditory information, for example, would be able to appreciate that sights and sounds can sometimes be traced to the same objects. In birds and mammals the evolution of interneuronal systems (i.e., brain structures) accelerated following the extinction of dinosaurs 65 million years ago. The most striking evolutionary event since the demise of the dinosaurs has been the development of the cerebral cortex, the largest and most complex part of the brain. And with the enlargement of the cerebral cortex we are brought further up the scale of conscious experience.

The substrate of consciousness is organized by natural selection. That is, the genes hold instructions for when, where, and how many neurons will be built. The genes further ordain much of the complicated neural circuitry. But as more interneuronal systems were brought into play there also emerged a potential for additional circuits to be constructed and reinforced by the accidents of individual experience. Add one neuron to the brain and you add the potential for many new interactions. Add millions and the possibilities for information events multiply endlessly. In other words, as brains became more complicated the organizing influence of "software" came to rival the organizing influence of "hardware." The strategy of encoding information in direct response to the environment mimics and bypasses the indirect and tedious logic of storing information in the genes. The ability to learn directly from the environment had enormous survival value. Organisms whose behavioral repertoire is mostly hardwired are unable to adjust their behavior and may therefore go extinct under conditions of sudden environmental change. But organisms that can design much of their behavior on site to accommodate environmental shifts will have a fighting chance. Learning is a splendid thing.

More neurons . . . more information . . . more learning . . . more memory . . . more versatility in behavior. The evolutionary advantages in this strategy should be obvious. But consider the eventual problems of carrying the idea too far. For one thing, there is a point of diminishing returns to the learning strategy. Any neural system organized exclusively by learning would take an eternity to acquire the programming minimally required for normal functioning. Moreover, such a system would be much too slow to respond efficiently in an emergency. A monkey who has to learn that tigers are dangerous is not likely to leave behind many descendants. And beyond these inefficiencies, consider the impediments to social existence that would result (bearing in mind that an exclusively learning-dependent species would of necessity be social). If all neural organization were left to the idiosyncrasies of experience then in a very brief period of time individual patterns of perception and behavior would diverge so far that social solidarity and cooperation would be out of the question. A society of radical eccentrics is no society at all. A certain measure of learning may be an adaptive thing, but too much of it results in inefficiencies and social chaos. From an evolutionary perspective the *tabula rasa* is a nonstarter.

Evolution has designed a solution for maximizing the advantages of learning while preventing the gross inefficiencies and the asocial drift of a tabula rasa. The solution is culture. Cultures maximize the power of learning by making its fruits heritable. Learned information encoded in the neural system of, say, a dog will be relevant to its own survival and well-being. But there are no means by which the dog's offspring might take advantage of the information. It is trapped in the old dog's brain and cannot be transmitted to the next generation. When old dogs die their tricks die with them. Humans, by contrast, are equipped to externalize acquired information by formulating symbols that can be decoded by other humans. The neural organization achieved by the learning of one individual may therefore be replicated in another by the process of symbolic communication. Culture consists of all the phenomena associated with this process of transmitting acquired information.

In recent years serious challenges have been put forward against the claim that culture is unique to humans. The facility of chimps to learn sign language, for example, in addition to many observations of nonhumans to learn by imitation have convinced many that the transmission of learned behavior is common in other species. Still, there remains a very large difference between the mediation of learning by imitation and mediation by symbols. Perhaps the former should be labeled "proto culture." The difference between proto culture and symbolic culture might be seen by analogy to the difference between the vascular information systems of plants and the neural systems of animals, the latter of these being far more efficient and versatile.

Another analogy—that between genetic evolution and cultural evolution—has occurred to many evolutionary thinkers. Richard Dawkins has even proposed the "meme" as a unit of cultural inheritance, analogous to the gene of biotic evolution. A meme is any meaning or behavior that may be formulated symbolically and decoded by a recipient. Cultures, therefore, may be seen as traditions that generate a diversity of memes and then select which of them will be preserved. In these terms we may see more clearly the logic of cultural evolution. By acts of creativity, synthesis, and even misunderstanding, cultures will generate a large diversity of new meanings, far more than could ever be preserved by a cultural tradition. There are more songs composed than recorded, more laws proposed than passed, more books written than published, more pictures painted than hung, and so on. One of the principal functions of a cultural tradition is to select out those meanings it considers unacceptable. Here we observe the Darwinian logic of variation followed by selective winnowing. The memes left intact by a cultural tradition will have a profound influence on the organization of consciousness. It is by virtue of the memes shared by participants in a cultural tradition that such traditions come to have their own distinctive identity. And in sharing a large number of memes these participants enhance their ability to achieve and maintain solidarity and cooperation.

The importance of story should not be overlooked here. The cultural narrative that integrates ideas about how things are and which things matter will be the ultimate standard against which any new memes are judged. Thus a story tradition functions as the principal agency of cultural selection. If a new idea is pronounced unfit by the custodians of the narrative tradition then measures will be taken to discourage it.

These dynamics help us to understand the process of divergence between cultural traditions. If a subgroup exercises the spread-out option and migrates to the next valley it may very well lose contact with the original group. There it will inevitably have its own unique experiences, giving rise to a totally new range of memes. Despite efforts to preserve the selective role of the original narrative it would be only a matter of time before new schools of interpretation would have a corrosive effect. New stories will emerge too. Different variations of memes together with different standards of selection would, in time, diminish the overlaps of neural organization to the point where social intercourse between the two groups would be fruitless. By such dynamics thousands of distinct cultural species have evolved during the course of human history. There are lots of different ways to be a human being.

But how many ways, and how different are they? These are important questions because they bear on the credibility of making universal assertions about human nature. If the range of cultural divergence is unlimited, and if cultures are in the most consequential ways incongruent, then we are left to conclude that there is no such thing as a common human nature. Are there many ways to be a human or are there different *kinds of humans?* The old nature-nurture debate again. Which of these is finally the more decisive: the neural organization wired in by the genes, or the neural organization programmed in by cultural memes? Is our behavioral apparatus so plastic that we might effectively transcend a common core of humanity? If so, then a search for the universal features of everybody's story as a basis for global solidarity and cooperation—that is, the project of this book—would appear to be a pipe dream. But if it appears that acquired characteristics are mere accidental variations standing atop a more consequential iceberg of inherited characteristics then it makes sense to acknowledge a universal human nature.

So which is it? The climate of informed opinion about these matters has shifted in recent decades. In the early decades of the twentieth century, during the ascendancy of the social sciences, it was fashionable to regard biological influences on human nature as trivial when compared with the overwhelming influence of cultural and individual factors. Nurture, it was believed, had the power to trump nature. Existentialist philosophy reinforced this perspective with the doctrine that individual human beings were uniquely and radically free to determine their own nature. To the extent that human nature exists it is completely relative to the choices made within the sovereign domain of personal existence. Since the midcentury emergence of neo-Darwinism, however, these claims for cultural and personal relativism have been challenged by a more biologically oriented view suggesting that while individual and cultural programming are significant they come far short of trumping the universal layers of organization ordained by the genes.

The most plausible story appears to be this: the accidental wisdom of natural selection has equipped the human genome to build a modular brain wherein fixed neural structures stand prepared to adapt, within limits, to particular variations in the environment. Nature builds the instrument and equips it with knobs for fine tuning that may then be adjusted to suit particular circumstances. Thus, for example, there are fixed language modules in the brain predisposing all normal humans to learn whatever particular language is presented to them by their cultural context. There are also modules that control perception, though these may become biased by the particular influences of personal experience and cultural traditions. And there are modules for initiating and regulating social interactions, which may be manipulated to some extent by local symbols and customs. Considerable evidence is mounting to suggest that these (and other) inherited modules are hierarchically organized to call into play a rich diversity of the brain's mechanisms for information processing. Especially interesting from the perspective of this book is the extent to which the modules comprising our cognitive systems are set up to acquire emotional markers. A central organizing feature of brain function seems to be an integration of information about how things are with information about which things matter. We may even expect neuroscientists to announce someday soon that they have identified story modules within the brain—that is, neural structures enabling the construction of coherent and emotionally textured narratives of ongoing experience.

There is a human nature, a wide range of universally endowed defaults and dispositions shared across the species. These universal characteristics are fixed in neural systems that are in turn constructed from information stored in genetic material. But many of these systems are open to modulation by acquired information. Sometimes we override our default behaviors by repressing them or by designing alternative behaviors. And sometimes we reinforce them with learning. Innate abilities, too, may or may not be developed by the refinements of learning. The point is that there exist many possibilities in human neural systems for variations of organization. Paradoxically, the potential for particular variation is a central feature of our universal and invariant human nature. Our most impressive constant is our potential for change. It is this potential, once it is differentially exploited by cultural traditions and personal ambitions, that makes it sometimes difficult to apprehend the deeper truth that all humans share a common story.

Interlude: The Sources and Splendors of Mind

Say something offensive to a flatworm and you will not get much in the way of a response. But offend one of your friends and there is no telling what to expect. I have been trying to stress that the difference here has everything to do with the organization of information-processing systems. The flatworm has all the basics—sensory devices, central nervous system, motor systems—but these are as simple as they can be. Whereas your friend's information systems are as complex as these things get. In simple reflex systems, much like the flatworm, the ratio of information input to information output is about one-to-one. But in the neural systems of a human being input information is amplified, analyzed, and enriched by stored information in ways that produce a ratio differing by many orders of magnitude. Somewhere along the continuum conscious experience begins to appear.

Speaking of offending one's friends, I have many friends whose emotional markers send up flares when they hear accounts of subjective existence cast in the vocabulary of input and output information. To speak of the life of the mind in terms of neural organization, they like to say, is to overlook what makes it special, that is, its internal *qualia.* And more to the point, the perspective of neurobiology leaves us with a rather inglorious view of the self, a self that is essentially reducible and derivative, not exactly a real person. How could anyone be inspired to a sense of gratitude for the lives we have when finally those lives amount to little more than precarious piles of excitable cells?

Perhaps. But consider this. A colleague once showed me a gift he received from one of his students. It was a jigsaw puzzle assembled by the student and mounted in an attractive frame. A nice gift, but to my mind my colleague seemed excessively grateful for it. After all, what does it take to put a puzzle into a picture frame? I remained unimpressed until I was informed that the student was blind. Sometimes the magic and the glory are less in the thing itself than in how it comes to be. And so it appears to be with the self. It took nearly four billion years of biotic evolution and a hundred thousand years of cultural evolution to organize selves who are capable of composing and comprehending this sentence, and neither you nor I had anything to do with it. For what it is worth, to be a self is to inherit a fortune of organization. Our part is merely to invest it.

We have the good fortune to be alive in an especially exciting period of intellectual history, a time when technological breakthroughs are allowing neuroscience to inch its way toward a more satisfying yet more puzzling view of the self. The emerging picture places the self somewhere between the realism of

Descartes and the nonrealism of Buddhist philosophy. Descartes believed that the self (or soul) was a unit of mental substance, an irreducible basic stuff that possessed powers of rationality and volition. For him, mind was enduring and indestructible, noncontingent and completely independent of the properties of matter. The self was derivative only in the sense that God was its ultimate creator. By contrast, in some versions of Buddhist thought the self is construed as a nonreality, a contrived fiction resulting ultimately from illusion. The self is thoroughly contingent and derivative, enduring only so long as the illusion endures. The point is to deconstruct the illusory construction.

The truth about the self appears to be somewhere between these views. Antonio Damasio speaks of the "neural self," which we might construe as an *emergent* reality. The neural self may be regarded as completely derivative and contingent, though it is derived from the properties of real things (neural systems), and not from a sad mistake. The neural self also endures, though only in the sense that it is continuously reconstructed by interactive neural systems. The self is not an illusion, but neither is it an independent reality. It is a tissue of imagery, an achievement of narrative ability.

This account seems incomplete. Since the self is an emergent reality the only sensible way to proceed is by describing the process of its emergence. Here follows a brief summary of Damasio's interpretation of the self. The body is equipped with cognitive systems, that is, sense organs and neural assemblies that transform stimuli into representational object images. The process of generating these images is ongoing, so that the brain is able to construct a fluid narrative of what appears to be going on in the surrounding environment. The body is also equipped to monitor its own states in roughly the way it monitors the external world. This monitoring results in the generation of representational body images (i.e., feelings and emotions), and a parallel narrative of ongoing events in the body. So far so good—two parallel narratives being continuously constructed from images produced in the brain. The next steps involve the brain's reactions to its own narrative constructions. Object images provoke a response in the organism. This "organism-in-response-to-object-imagery" is yet a third set of events that the brain is equipped to monitor and to construct into narrative form. So now we have three interactive narrative streams: a narrative of what is going on "out there," a narrative of what is going on "in here," and a metanarrative of the perceiving-responding organism. From this third narrative, Damasio believes, there arises the reality of self, *a unique subjective perspective* that corresponds to a sense of being alive in an object-filled world.

This speculative story presents us with a rudimentary self, the sort of subjective reality that might be enjoyed, more or less, by species of birds and lower mammals. A basal subjectivity. But as new modules of the brain come on board the rheostat of consciousness brightens. Among primates the images brought into the narratives would be more complex, featuring greater subtleties of pattern, longer persistence in memory, more refined motor imagery, and, of course, a broader range of emotional textures. Narrative plots thicken. Add language modules and the narratives open into a longer, collective past and a more distant, planned future. But more important, language mediation broadens the scope and streamlines the process of monitoring the imagined perspectives of other neural selves. Narratives now have an intersubjective dimension. Shared hopes and fears become integral parts of the emotional landscape. The articulate story traditions of a culture interact deeply with the narrative streams constructed by individuals. Thus the neural self emerges as a convolution of individual and social construction.

This is how we come to have the lives we do, by a further organization of matter into more complex systems. A person, a self, is a natural system designed by the cumulative achievements of, first, the course of biological evolution; then the course of cultural evolution; and finally the course of personal development. To the minutiae of these often wayward processes we owe all that we are. The neural self we encounter in this picture is precarious and fragile, vulnerable to disorganization by injury, disease, aging, and abuse. Sometimes even the most trifling affairs—a mild rejection, a simple mistake, a slight misunderstanding—can trigger a series of events resulting in personal chaos. But the potential is also there for transcendent flights of imagination and understanding. The neural self can escape the limitations of space and time to create images belonging to the past and the future, to behold real but unobserved objects, events, and relations unimagined by selves of less endowed species. The neural self can detect complex problems and invent insightful solutions. And it can transcend even the perspective of its own making to commune with other neural selves. Such is the power of organizing neurons.

But the neural self is not infallible. Like the mechanisms governing the genetic code, our neural systems routinely falter. We forget, we overlook, we form bad habits of mind, we rush to judgment, we misperceive, we persist with poor solutions, we are vulnerable to deception, and we excel at blunders of communication. Yet these unfortunate possibilities are all part of the territory of the neural self, they are the price we pay for the potential triumphs of mind.

There are undeniable insights in the Buddhist and Cartesian conceptions of the self. The Buddhist doctrine rightly sees that the self is a contingent construction that emerges only to make excessive claims for its own importance. And the Cartesian doctrine rightly insists that the self is real and central to coherent human existence. But these doctrines are finally unsatisfying for the most obvious reasons. The Buddhist doctrine denies an undeniable reality, and it counts for nothing that it does so for perfectly honorable moral purposes. And Cartesian dualism contradicts the plain fact that every aspect of our mental lives is derived ultimately from the properties of matter. The self is real but it is not Cartesian, it is neural. The neural self, as we have seen, affirms all that is sensible in these alternative doctrines yet avoids the absurdities.

There may be an element of species arrogance embedded in the attempt to exalt the power and the glory of human conscious existence. If there is, I hasten to confess to it, for I believe that the human neural self is the most thankworthy reality in the universe. If I could save but one thing it would be this. The

Buddhist doctrine of salvation undermines such arrogance by declaring that the self is something to be saved *from.* And the Cartesian salvation scheme trumpets the arrogance by negotiating passage of the soul into the realm of the gods. But once again the neural self gives us a different way to think, beginning with the recognition that saving one thing is never an option. Perhaps this fact can help us to see an element of constructive moral value in our species arrogance. Human conscious existence is the most highly contingent reality we know—it is the *last* thing we can save. If we cherish our existence, if we are thankful for the splendors of consciousness, we are therefore bound to value all that is presumed by it. If we want to save the neural self then we will fiercely safeguard the integrity of the natural and social systems by which it comes to be.

Part II

WHICH THINGS MATTER

Chapter 4

What Matters Ultimately?

Nothing matters to a block of ice. It is completely indifferent to external conditions, equally satisfied to be a pool of liquid or a steamy mist. Likewise, nothing matters to a molecule of H_2O. If agents in the environment break down the molecule to involve its atoms in other affairs it is just as well. Physical systems are like that: they just do not care, they cannot be bothered. Living systems, by contrast, *do* care. They have interests, and having interests they are not indifferent to the conditions in which they find themselves. Some conditions are preferred over others, namely, those conducive to the continued integrity of the organism. Can one say the same for a block of ice? As temperatures rise would it make sense to say that the block of ice now *prefers* to be a pool of liquid? Of course not, but *why* not? Our answer must be that there are no signs that water thrives any better in a liquid state than in a solid state; nor does the block of ice show signs of seeking out any set of conditions over any other set of conditions. It is all the same to a block of ice.

We move from a description of how things are to a discussion of which things matter with the observation that some things just *are* such that certain conditions *matter* to them. Caring, having interests, being programmed to prefer certain conditions, seeking out and thriving under those conditions—these are all ways to speak of an evaluative property emergent in living systems. In short, motivation can be claimed as a universal characteristic of life. Living things are motivated to seek out and/or create the conditions perceived to be favorable to their own thriving. Life is motivated matter.

This chapter has a very simple thesis. It is that for all species of life there is one common thing that ultimately matters, that is, living. Yet, this is far from obvious. Algae, for example, may be said to have no appreciation for anything as abstract as life, and thus could not possibly value it. But certainly the ability to appreciate abstractions is not the same thing as embodying values. Algae do not appreciate the idea of sunlight either, nevertheless they seek out sunlight and thrive on it. Moreover, an alga's preference for sunlight is instrumental to its viability, it lives in part by *virtue* of its valuing sunlight. To the alga, we may therefore say, sunlight is valued as a means to the more ultimate value of viability. Of course, if an alga could speak it might well say it cares only for sunlight and does not give a fig for this thing we call life. Here we might be tempted to say the alga got it only half right: we value in order to live no less than we live in order to value. Dead algae do not value sunlight. Dead algae cannot be bothered.

If all we had to go on were the example of algae then we might conclude that sunlight is the ultimate value. Indeed, we might say that the value of sunlight is identical to the value of life. But there are other living things having different specific vital interests that function in an equivalent instrumental way. Nectar is to bees what sunlight is to algae, we might say. And what is that? Well, *means* to viability. The specific means have the value they do relative to the ultimate value of living. Presumably, if algae and bees were pressured to find alternative means (say, artificial lighting or fruit punch), they would value these as they valued the original means.

And so it is wherever we look in the realm of living things. There is always a set of particular conditions valued by organisms of the species that, when pursued effectively, deliver the ultimate good—that is, life. Wherever life exists the conditions for living are valued. To say that viability is a universal value embodied in all living things is merely to generalize to the level of principle what is always and everywhere observed.

Always and everywhere? Are there no exceptions? In the way we have been proceeding it would seem right to conclude that the value of viability is both universal and objective. In fact, I confess to a high level of comfort with both of these claims. Yet neither of them finds favor among the holders of informed opinion these days, so a defense may be in order. Before defending these claims about the value of viability, however, I should make an attempt to clarify what I mean by it. By viability I mean something very close to the literal sense, that is, "likely or able to live." To value viability is to value the continuation and fulfillment of life, values that merge with the process of diversifying the forms of life. Viability and biodiversity are indistinguishable values. Once life appears in a nonhomogeneous environment then, as we have seen, its continuation and fulfillment inevitably results in a diversity of forms. We may also say that the ultimate value for the continuation-fulfillment of life transcends the distinction between survival and reproduction. For our purposes the value of viability may be seen to include the values expressed in the struggle of individuals of all species to survive as well as the values expressed in efforts of all species to achieve reproductive success.

Now to proceed with the business at hand, which is to defend the claims that viability is both a universal and objective value. The claim for universality says that all species are designed to optimize viability above all other valued outcomes. And the claim for objectivity says that all behaviors to the contrary may be judged with absolute confidence to be wrong.

Those intent on denying the claim for universality will draw attention to alleged exceptions among humans. Here may be found examples taken from religious traditions claiming that the true meaning of life is not in living but rather in the transcendence of life. In Christian tradition, for example, followers are encouraged to care less for the prospect of life than for the prospects after death. The life one has in the natural order has mere instrumental value as a means to a more ultimate form of supernatural existence. In itself, viability is hollow and without enduring promise. In this example—and there are many others—we appear to have decisive evidence against the claim that viability is universally valued.

But things are not always as they appear. It is important not to confuse expressions of "the meaning of life" with the embodied value of viability. When individuals articulate the meaning of life they are attempting to specify *why they value life*. Those inspired with Christian hope may insist that life has no meaning or value apart from the prospect of transcending it. On the surface this may be seen as a rejection of the ultimate value of viability, but in fact it may be only one of many particular ways to affirm viability. For example, it is common for individuals to claim that life has no meaning apart from their families, friends, work, hobbies, or even their pets. And it is true that the loss of one's job or a loved one (or the loss of religious faith) may pitch a life into despair, but in the course of time persons normally find a way to reconstruct meaning in their lives. Therapists tell us that this process of reconstructing the meaning of life is very common. But it is hardly a process that could be motivated by the meanings lost. There may indeed be a radical pluralism of reasons given for valuing life, but this is no evidence against a universally embodied value that functions as the ultimate source of these diverse reasons.

One further note about religion. Whatever religious memes appear to say on the surface, it is reasonable to view them as devices designed for optimizing the continuation of life. That is, these meanings normally function as instruments for generating the kinds of emotions and behaviors that are adaptive in the evolutionary sense. This is precisely what we should expect once we admit that culture itself is a complex biological adaptation—that is, a collection of shared meanings that assist our species in meeting the challenges to viability. Religious systems, as cultural artifacts, have emerged because they have proven to be biologically significant. Thus, despite appearances to the contrary, religious memes exist in service to the ultimate value of viability.

Next comes the defense of moral objectivity. I have said that any behaviors contrary to the value of viability may be judged with objective certainty to be wrong. That is to say, the standard of viability is what *makes* something right or wrong. If an act or rule is wrong it is *because* it compromises the value of viability, and if an act or rule is right it is *because* it promotes the value of viability. Bold and unambiguous claims of this sort have the appearance of simplifying the moral life, but nothing could be further from the truth. Having a crisp and objective moral standard does not mean that particular moral judgments will themselves be objective, or even easy. Compare the standard of viability with the divine command standard. Divine command theory says that any act going contrary to God's will is objectively wrong. As straightforward as this sounds, theists are often left to their own subjective devices in deciding what God's will actually is. The standard of viability is no better off. Even if we agree that viability is an objective value there will remain plenty of room for disagreement about which behaviors comply with it. The moral life is always messy.

But can it be agreed that viability is an objective value? Anyone making such a claim for the objectivity of any value must sooner or later do battle with the infamous *naturalistic fallacy*. David Hume, in the eighteenth century, and G. E. Moore in the twentieth, warned us of the dangers of this fallacy. It is rationally illegitimate, they insisted, to derive value claims from objective factual claims. Facts (i.e., accounts of how things are) constitute one sort of thing, but values (i.e., appraisals of which things matter) constitute quite another sort—and the two are not commensurable. No matter how deeply one is committed to a moral value, there is no way to establish the value by a demonstration of facts alone. There must be at least one given (or assumed) value. To illustrate, we cannot infer *from* "a high-fat diet is bad for one's health" *to* "one should not eat a high-fat diet." The inference requires an intervening value assumption, such as, "one should not jeopardize one's health." Once the intervening assumption is allowed, the inference can go forward. But facts alone cannot establish the objectivity of values, any more than wishing for something makes it come true. Ever since Hume and Moore the prevailing view among moral philosophers has been that there are no demonstrably objective values. This suggests that one may take a moral stand against the value of viability without fear of inconsistency.

At this point I need to make one thing clear. In most cases (all but one, actually) I agree with those who cite the naturalistic fallacy in their opposition to the objectivity of values. And I make a point of deferring to no one when it comes to fearing moral dogmatism. But in the case of viability there is good reason to make an exception. In other words, I believe we are justified in asserting viability as the one and only objective moral standard.

Here are my reasons. I begin with the most basic rule of morality: *ought implies can*. This rule says that one cannot reasonably lay moral obligations on persons to do what is not in their power to do. For example, you cannot blame people for not photosynthesizing because photosynthesis is not a human option. By the same rule we do not place moral sanctions on infants when they soil their pants. They cannot help it. In other words, moral judgments make sense only where behavioral options are possible, that is, where information-processing systems are sufficiently plastic to enable individuals to override default behaviors in favor of learned (and valued) behaviors. So far so good. This means that only a small group of species might qualify as moral agents, and then within this small group we make numerous exemptions where limitations on behavior warrant them. Thus, for example, I might reasonably shame my dog for soiling the carpet but not for failing to go to college.

Now to come to the heart of the matter. It seems clear to me that the entire domain of moral reasoning depends on there being in the world something we might call moral consciousness, that is, the ability to generate a variety of options for behavior together with the ability to winnow them selectively on the basis of moral rules. Apart from these abilities nothing in morality makes sense. This said, we may now ask about the value of these abilities. If moral consciousness is a good thing then what makes it a good thing? The biological answer is that moral consciousness is a good thing because of its adaptive value, that is, its power to assist us in meeting the challenges to viability. This means that the value of morality *per se* is derived from the value of viability. But if the value of having a range of behavioral options is relative to viability then the value assigned to any particular option is also determined relative to the standard of viability, and if the value of every moral option is relative to the same standard then that standard may be said to be objective. If one therefore chooses to act in ways that contradict the value of viability one thereby chooses contrary to the conditions of one's choosing. I believe this qualifies as an objectively irrational act.

There is nothing to be gained by sinking further into abstract philosophical disputes. The important point is that the value of viability deserves special status as a nonnegotiable part of everybody's story. It is the one value to command all others. Indeed, I have given sufficient reasons for thinking it is irrational to affirm any value as more binding or less optional than this one. To sum up: it is wrong always, everywhere, and for anyone, to value anything above the enduring prospect of life.

One final qualification. The above argument makes no claim to having "cracked" the naturalistic fallacy. It is still possible, in a strict logical sense, to deny the value of viability without contradiction. Nevertheless, the reasoning of this brief chapter does take us on a legitimate course *around* the naturalistic fallacy to the inference that *if anything matters then what matters most is viability*. And with this inference we may jubilate over the restoration of a form of moral foundationalism. That is to say, the only reasonable alternative to affirming the ultimate value of viability is to take up the nihilistic view that nothing at all matters. Nihilism is intellectually plausible enough, but it is so far from being existentially feasible that we may excuse ourselves for failing to nod as we pass it by.

When it comes to matters of human nature humility may be our best source of insight. Or such we might conclude from the lesson forced upon the Western mind by Charles Darwin. Before Darwin we had the luxury of basking in the glow of metaphysical delusions about the importance of our species. There we were, fixed in the great chain of being, halfway between the beasts of nature and the spirits of heaven. Before Darwin it was convenient to consider ourselves the whole point of the universe, created in God's own image as the special object of his love. But since Darwin we have been forced to accept a more humble place in the order of things. Humans are but one species among several million, and we are no more the point of the universe than were hydrogen atoms a million years after the big bang, or the dinosaurs sixty-five million years ago, or even the Spotted Owl of the present era.

The humility foisted on us by Darwin has not been easy to accept, but we are finally getting there. And with a more humble perspective on ourselves we are better able to understand and to affirm the essential solidarity of all forms of life. We may be organized in different ways but all species of life are informed by the same genetic vocabulary. We may have specialized niches but we all share the same biosphere. We may compete and feed off one another but we do so with the same biotic license. As we discipline ourselves to take a wider view we begin to appreciate that the overlaps among species are much more profound and important than the differences. From the outer space of a Darwinian perspective life is a unity, a community of shared interest in the conditions of viability, apart from which there is no enduring promise. The driving theme of everybody's story is to understand these ultimate conditions and to value them ultimately.

Finally, dare we ask about the value of having ultimate values? For all this lofty talk about the value of viability, what difference does it really make in the actual business of living? Even if it is granted that the prospect of continued life is an objective and ultimate value for all life forms the fact remains that it is hardly ever recognized as such. Certainly not by cuttlefish and cockatoos. And even among humans the recognition of viability as the ultimate value is very rare. That is because viability per se is far too generalized a goal to be useful in the struggle for survival. How many of us wake up in the morning with agendas aimed at keeping our rate of chemical composition ahead of the rate of decomposition? That is not the way it works. Instead, we wake up with more interesting and immediate concerns in mind. We wake up from our night dreams only to take up our daydreams of love and hope and treasure-seeking—these are the concerns that drive us from day to day. Winning the war may be the supreme goal but, as every general knows, all the action takes place at the level of particular battles. So it turns out that the ultimate value is often or even normally obscured by more specific and mundane interests. And wisely so. Generals who try to win wars generally lose them. Nevertheless, it makes occasional good sense to focus on the ultimate goal as a way to restore our perspective on immediate concerns, and also as a means for assessing the effectiveness of strategic plans and tactical maneuvers. Sometimes it is important to crawl out of the trenches and consult the big picture, so that one is prepared to make necessary adjustments. Who knows?—doing so might result in choosing different battles.

Chapter 5

What Matters Proximately?

Suppose you are in perfect health and a team of doctors approaches you offering a free ride on medical life-support systems for the rest of your life. There is a surplus of equipment, they explain, and the hospital has decided to make the benefits available to the winner of a lottery. You are the lucky one! They go on to explain that the life-support systems will extend your life far beyond the most optimistic expectations. You would never have to work again, never worry about accidents or health-care problems, and never fret about where your next meal was coming from. They are offering you a guarantee of viability, absolutely free.

Would you accept the offer? Certainly not, but *why* not? If it is truly the case that viability is the ultimate value for all living things then it would appear senseless to refuse. Suppose you decline the offer by saying you have not had any children yet and feel compelled to enhance the prospects of life by reproducing. No problem! The doctors promise to extract gametes from your body and mate them for you in the lab—as many as you like. When the children are "born" they will be placed with proper, loving families. Not to worry.

Still a nonstarter? But why? Because, you say, the life offered here would not be a life worth living, it would be empty of all satisfaction. The deeper reason for this is that you come equipped with your own life-support mechanisms, your own strategies for answering the challenges to viability. And the only tangible sense you have for the value of life is derived from these mechanisms. The offer might be acceptable if you did not have these systems, but you *do* have them, and in having them you cannot be drawn to the prospects of a life without them. Your life-support strategies have the important feature of being governed by motivational systems, that is, systems that arouse you to design behaviors resulting in their satisfaction. It is by having these systems that you come to behave in ways that keep you alive, but it is also by experiencing the goods internal to these systems that you come to have a sense that life is worth living. The same goes for other species. Indeed, one of the most useful approaches to the differences between species is to examine their systems of motivation. By exploring the details of these systems we gain insight into what matters to a particular species. If viability matters ultimately to all forms of life, then we might say that *proximate values* are those embedded within the general strategies of a species to achieve viability.

Motivational Systems

If we want to know what matters proximately to human beings, we should try to get a fix on human motivational systems. These systems are numerous and varied, full of details that would take us far afield. Humans are as complex as they are precisely because they have so many motivational systems operating simultaneously. For our purposes it will be enough to gather these systems into three broad categories: *curiosity motivators, hedonic motivators,* and *social motivators*. What matters to humans is to maximize the goods inherent in satisfying these motivational systems.

Nothing could be more misleading than the old adage that curiosity killed the cat. Better to say that curiosity keeps cats alive. For cats, like all other animals graced with central nervous systems, are well served by mechanisms inducing them to explore their environments for useful information. I am here using the term "curiosity" to include the full range of cognitive and sensory arousal systems, those systems predisposing us to monitor the environment for novelty and change. These systems function to integrate a wide diversity of information into larger frameworks of meaning. Curiosity is a deficit state which, when aroused, will normally set an organism on a course of action designed to repair the deficit. Consider the state of someone aroused from sleep by a loud noise just outside the bedroom window. Persons aroused in this manner may rush to the window and throw open the sash to see what is the matter. Or they may lie still, with bated breath, awaiting further sensory input. These are slightly different strategies but they amount to the same thing—that is, behaviors designed to gather information until some desired satisfaction takes place, thereby restoring the motivational system to equilibrium.

It would not help in this case to run to the pantry for a snack, or to brush one's teeth. Such ill-designed responses would not address the deficit. Curiosity is a cognitive deficit that can be repaired only by an experience we might describe as recognition, understanding, or intelligibility. But the phenomena associated with this motivator are varied. For example, curiosity may be aroused by acts of reflection or memory events as well as by sensory stimulation. And the nature of problems may range all the way from simple tasks of perception to complex problems involving high-level abstractions. Whether one is trying to figure out the source of noise in the garden or the most elegant proof for a theorem of mathematics, the logic of motivation is the same: a challenge to understanding keeps the individual aroused in a deficit state until appropriate behaviors result in the experience of intelligibility. If curiosity is not satisfied in due course, then an individual might invent an explanation (e.g.,

"It must have been the wind"), or find a convenient way to reduce stress by disengaging one's interest (e.g., "I've got better things to do").

The purpose here is to discover what matters to human beings. The suggestion has been that we may gain insight into human values by looking at motivational systems, among which is found a wide range of curiosity-provoking systems. These systems motivate us to seek satisfactions in the form of experiences of intelligibility. From this we may judge that intelligibility matters deeply to human beings. It is one of the primary values guiding our being in the world. As Aristotle observed, all humans desire to know, and when this desire is satisfied we behold it as a realization of intrinsic value.

There are more ways than one to sustain a cat. Curiosity helps, but so does the pleasure principle. A second set of arousal systems may be grouped together as the hedonic motivators, systems that become activated by certain changes within the body. When changes occur in metabolism, hormone levels, body temperature, and other regulating systems, individuals enter deficit affective states that seek repair by means of goal-directed behavior. If I am deprived of water I will eventually become aroused by thirst, whereupon I will be hedonically motivated to seek the means of satisfaction. If I stand outdoors in the rain I may take a chill and become motivated to seek repair in shelter. If I am exposed to erotic images I may become sexually aroused to seek a mate. And so on. A good share of our wakeful time each day is spent on behaviors designed to repair deficit states aroused by hedonic subsystems. Like cognitive systems, hedonic motivators have their appropriate satisfiers. If I am thirsty it will not help to arrive at some profound insights about the nature of thirst. It is not that kind of deficit. My thirst will keep me aroused until I have the pleasurable experience that comes with taking a drink. Pleasures (and relief from pains) are like intelligibility—they are systemic satisfiers, valued for their own sake because they are goods internal to human motivational systems.

All things considered, it should be enough for an animal to be equipped with curiosity motivators and hedonic motivators. The first of these arouse the individual to acquire a working knowledge of the external world, which it then uses to secure means for satisfying the second. What more could an organism ask for? Well, a bit of help might be nice, especially if (or necessarily if) you happen to belong to a social species. For many species, the two vital operations of knowledge and exploitation are undertaken as collective enterprises, more or less. Less, perhaps, if you are a smugly independent cat, but very much the more if you happen to be a dolphin or a human. Highly social species depend on strategies for communicating information and consolidating labor, both of which require solidarity and cooperation. Some of these species—the ants, for example—engage in social behavior at the behest of reflex systems, but in large animals social behaviors are mediated by emotional arousal systems, that is, the social motivators.

Among the social emotions are the following:

disappointment—sorrow—grief—despair
affection—adoration—love—devotion
resentment—outrage—hatred
regret—shame—guilt—remorse
humility—respect—reverence—awe
pity—sympathy—empathy—compassion
envy—jealousy
gratitude—obligation—duty

Each of these emotions represents a deficit state that may be aroused as individuals assess patterns of social interaction and social signaling. Once aroused, individuals remain in a deficit state until some behavior undertaken or some circumstance observed functions as a satisfier. If deficits are left unrepaired individuals may advance in frustration to more intense deficit states, which may then seek more exaggerated forms of satisfaction. It is very difficult to be precise about the dynamics of satisfying the social emotions because of all the variables introduced by psychological idiosyncrasies and social conventions. But a few general remarks may help to illustrate how these systems work—sorrow and grief seek repair through overt expressions inviting the sympathetic attention of others; affection seeks repair through behaviors designed to achieve social bonding; reverence seeks repair in the performance of submissive behavior; gratitude seeks repair through reciprocal behaviors; guilt seeks repair through behaviors soliciting signals of forgiveness; resentment seeks repair in some form of restitution; sympathy is satisfied by behaviors designed to rescue or console.

These arousal systems we call the social motivators because they are responsible for instigating socially constructive behaviors among otherwise self-serving individuals. The result of all this emotional disturbance and repair is to create extensive overlaps of self-interest. When one individual is emotionally aroused to help another they both gain, the benefactor by satisfying an emotional need and the beneficiary from the assistance. Social emotions are shuttles in a loom of social exchange, by their arousal and satisfaction there is woven a social fabric of solidarity and cooperation.

Perhaps the most consequential of all the social emotions is one not listed above, namely, the self-esteem motivator. The self-esteem motivator qualifies as a social arousal system because the system itself emerges in the process of socialization. As children acquire a socially induced set of norms for behavior they enter into a lifelong process of continuous self-monitoring. That is, they begin to measure their own performances against socially objective standards of behavior. Whenever the self-monitoring process turns up evidence of shortcomings the individual becomes emotionally aroused by a sense of deficient self-worth, and this deficit seeks repair through behaviors designed to merit attributions of positive self-worth. I have distinguished the self-esteem motivator from the other social emotions because, as we shall see, this arousal system enters into the game as a wild card.

There is, of course, much more to say about the details of the social emotions, and there is even more yet to learn about them. One of the most interesting questions about these systems is the extent to which they operate in nonhumans. Another domain of interest has to do with interactions among these social emotions as well as their interactions with the curiosity motivators and the hedonic motivators. But enticing as these issues are, they must not tempt us away from the main point, which is to find out which things matter to human beings.

We got started into this inquiry on the premise that an exploration of motivational systems would yield insight about human values. The supposition was that what matters to humans is to achieve the goods internal to these motivators. So what have we learned? The story seems to be that humans are aroused by motivational systems of three basic types: curiosity motivators arousing us to render our experience intelligible, hedonic motivators arousing us to seek pleasure and to avoid pain, and social motivators arousing us to seek emotional fulfillment. Intelligibility, pleasure, and emotional fulfillment—these things matter to us as intrinsic goods because they satisfy our longings.

They do indeed, but there are some problems with this story. To begin with, there appears to be an intuitive problem. The discussion so far seems to imply that whatever satisfies an urge is deemed good by definition. This follows from the view that satisfiers are goods internal to motivating systems. On one level this seems perfectly true. That is, if we limit our judgment to a particular arousal—say, an itch—then relative to the itch a satisfying scratch would be, by definition, good. Likewise, when I am aroused by hunger a satisfying sandwich would deliver an intrinsic good. Taken in isolation, arousal by arousal, there seems nothing wrong with the claim that satisfiers are intrinsic goods. But this piecemeal way of thinking would commit us to declaring the inherent goodness of many outcomes that strike us intuitively as very bad. The trouble is that deficit states and their satisfiers hardly ever exist in isolation. Let us say I get hungry and snatch away your sandwich to satisfy my urge. From the point of view of my hedonic motivator an intrinsic good has been served. But there *you* are, freshly aroused in resentment of my audacity, and very likely predisposed to give me a good thumping—which, relative to your social motivator, would also deliver an intrinsic good. It seems obvious that what is intrinsically good relative to an isolated context may be considered quite evil relative to a more inclusive one. It seems therefore wrong to say that our species might be well served by a general strategy that values the satisfaction of each and every urge toward intelligibility, pleasure, and emotional fulfillment. In the inclusive context of social existence some of these intrinsic goods would be destructive.

Consider another problem appearing at the level of individuals. The broad range of motivational subsystems simultaneously pressing their demands on individuals sets the scene for deep conflicts. What happens, for example, when an individual is aroused by curiosity in two directions at once? Say I am at my desk struggling with a mathematical proof and I hear a strange noise outside my office window. Obviously I cannot pursue both investigations at once. So how do I choose? Here is a case of psychological conflict that must be resolved. The value of intelligibility is of little help here since it is the prize of both pursuits. Or what if I am on the brink of discovery and suddenly hunger strikes. Here is an intrapsychic contest between intelligibility and pleasure. By what rules do I resolve it? Or what if I get caught between two social commitments? Or between a social commitment and a much needed rest? The point is that a person is a bundle of motivational systems and subsystems, all of which are capable of commanding the attention of the entire organism. It is true that we value intelligibility, pleasure, and emotional fulfillment as intrinsic goods, but it is also true that an individual cannot be all things to all motivators. Sometimes we are forced to sacrifice some goods in favor of others. But how? By what standards? The practical demands for resolving social and psychological conflicts suggest there is something to be learned about human values by addressing these questions.

I will propose a pair of mutually depending, yet mutually contending, proximate values: *personal wholeness* and *social coherence*. The claim will be that the human strategy for winning the war of survival is to fight on two fronts, the battle against personal disintegration and the battle against social chaos. If our species can manage to create the conditions for organizing well-integrated, stable, robust personalities, and if we can simultaneously create the conditions for constructing coherent, cooperative, harmonious groups, then we will have a good chance of defeating the odds against human viability. On the one hand, personal wholeness and social coherence are interdependent values in the sense that each is a necessary condition for the other. Whole persons cannot be nurtured in a context of social chaos, nor can a coherent social order be constructed by dysfunctional individuals. On the other hand, personal wholeness and social coherence are contenders in the sense that they threaten to undermine one another. Individuals tend to lay excessive demands on social resources, while groups tend to make excessive demands for personal sacrifice.

As we have seen, a person is (at least) a complex set of motivational subsystems, each with the potential to arouse the individual in service to its needs. The persistence and plurality of these demanding motivators make it essential that the individual suppress some of them in favor of satisfying others. Imagine what life would be like without the means for making such selections. You would not know which way to turn. You would try to do everything at once, thereby spoiling your chances of doing anything at all. You would, in very short order, go crazy. One of the ironies of human existence is that sacrificing certain intrinsic goods in the name of personal integration is a necessary means to maximizing such goods. Trying to have it all is a recipe for having nothing. Whole persons are those whose motivational systems are robust yet effectively managed—persons who are able to construct agendas of sequential tasks, to anticipate outcomes, to assign priorities, and then to attend to more important matters while momentarily suppressing the demands of competing impulses. By such means the whole person is able to achieve a state of functional unity against the odds generated by a plurality of motivational systems. "Whole" means both full and one, complex yet integrated. Our entire lives are constructed, moment by moment, through our efforts to harmonize the demands put upon us by our relentless needs for intelligibility, pleasure, and emotional fulfillment. By harmonizing these demands we maximize our achievement of goods inherent in motivational systems. Personality, then, is a metalevel achievement—an emergent, transcendent pearl of great price, purchased at the expense of many intrinsic goods forsaken. What matters to human beings is to become whole persons.

There are many different ways to be a whole person. A single formula for answering the challenge of personal integrity does not exist, nor could there ever be one. To suppose so would be to ignore the plain fact that each person's experience is unique. Your inherited combination of temperament, sensitivities, limitations, and abilities is not mine; and my set of environmental obstacles, opportunities, misfortunes, and advantages is not yours. Each individual is a lonely junction of these variables. Everyone has their own story emerging from a unique set of motivational challenges. The potential for human diversity is enough to defy all norms. And yet norms there must be, for without them we forsake the benefits of social existence.

The challenge of social organization is not unlike the challenge of multicellularity, that is, how to get a plurality of discrete parts to behave as a single entity. The answer is to get the parts to conform to certain rules, to maximize conformity of behavior among members of a group. There may be many authentic ways to become a whole person, but encouraging a diversity of these ways is hardly the way to create a coherent society. A coherent society is achieved by reducing diversity, by establishing normalcy—that is, by getting the wholeness process as close to a single formula as possible. Social coherence is not incompatible with personal wholeness per se, it is just incompatible with too many ways of seeking it. If there is too much diversity then social pressures will arise to standardize the pursuit of happiness. We know the lines: "When in Rome, do as the Romans do"; or, "Not as long as you're living under *this* roof, young lady!" Social order matters. Personal wholeness is to be encouraged, sure enough, but not in ways perceived to jeopardize social coherence. These values must be made commensurate.

Enter the wild card of self-esteem. The self-esteem motivator is the dominant organizer of the process of seeking personal wholeness. This is made clear by the extremes to which individuals will go to elicit signals of social approval. The self-esteem motivator is a wild card because it can be linked to almost any kind of satisfier, even self-destructive behavior. For example, if you could manipulate people into linking self-worth to blindness, then they will be motivated to blind themselves. A social group will maximize solidarity and cooperation by getting people to link their self-esteem to behavioral norms. When self-esteem is thus linked to social sacrifice, then individuals will be motivated to behave with a unity of purpose.

Of course the system does not always work to perfection. On the one hand, if socially induced norms become too constrictive they will obstruct attempts to satisfy some motives essential to personal wholeness. On the other hand, if social conventions are too permissive then conflicts of interest will begin to erode the social order. Every society, therefore, will have to negotiate compromises between personal ambitions and social responsibilities. Indeed, apart from such compromises there would be very little personal wholeness or social coherence. There is, of course, no absolute formula for working out these compromises, which accounts for the plurality of cultural traditions in the world. To the extent that cultural traditions exist in isolation from one another they will work out distinctive patterns of compromise. And when disparate cultures converge—as they did in the Axial Age, and as they are doing now—then traditional strategies for achieving personal wholeness and social coherence will begin to falter, calling for new means of compromise. The essential point to be seen here is that while the particular strategies for achieving these twin values are locally negotiable (and therefore culturally relative), the general strategy of having the values is not. In other words, wherever you find human beings you find a concern for the metavalues of personal wholeness and social coherence. To balance these values is our species' way of beating the odds against our survival.

And now to run up a subtotal of everybody's story so far told. In Part I I presented a very broad sketch of the scientific account of how things are in the world, including an account of human origins and human nature. This account qualifies for inclusion in everybody's story because it is the least contingent and the most accurate cosmology available to us. Everybody should embrace this account because it brings us as close as we have been able to come to the truth about how things are. In Part II I have been trying to say which things matter. Here the story is that nothing matters if there is no life—that is, valuation itself is an exclusive and essential function of living systems. Everybody's story includes the ultimate value of viability and the proximate values of personal wholeness and social coherence. Everybody has an interest in viability by virtue of being alive. This value we share with all living things. Furthermore, everybody has interests in personal wholeness and social coherence by virtue of being human, for these values reflect specieswide interests.

Ecotherapy, Psychotherapy, Politics

I have said that personal wholeness and social coherence are mutually depending and mutually contending values, that is, each enables and constrains the other. To these we are now obliged to add a third set of specieswide proximate values—those expressing the physical and biological conditions for human existence. Human beings work out their quest for personal wholeness and social coherence in the context of natural systems that themselves enable and constrain human life. Implicit in our valuing human life is a commitment to safeguarding—to the extent that we are able—the integrity of those natural systems that make such a life possible. All humans have a demonstrable interest in biospheric integrity—that is, an interest in sustaining indefinitely a level of biodiversity conducive to the pursuit of personal wholeness and social coherence. Biospheric integrity therefore joins personal wholeness and social coherence to complete our story of which things matter proximately.

The point I am driving toward is that the proximate values of everybody's story may be gathered under three principal categories of moral concern: *ecotherapy, psychotherapy,* and *politics.* The imperative of ecotherapy is to foster the conditions for biospheric integrity, that is, to act in ways designed to maximize biodiversity. Psychotherapy is governed by the imperative to act in ways that enhance the abilities of persons to achieve wholeness, thereby to maximize the goods inherent in motivational systems. Politics embodies the imperative to conform to social norms, thereby to maximize social solidarity and cooperation. These three domains of moral concern may be seen to overlap, as shown in Figure 3 (see page 122). If an act or rule for behavior may be judged to advance any of the three imperatives then it may be represented as a dot within one of the three spheres. The central point in all of this is to suggest that what matters proximately is to create the conditions for biospheric integrity, personal wholeness, and social coherence. If we can manage to organize ourselves in ways that satisfy all three imperatives then we shall have a fighting chance against the challenges of the global problematique. The key is to get as much human behavior as possible to land within the shaded area common to all three domains of moral concern.

In the introduction I claimed that the principal sources of the global problematique are overpopulation and overconsumption. These human excesses have already severely compromised biodiversity, perhaps to the point of triggering a mass extinction. The implications for ecotherapy are clear: serious measures must be taken to reduce human impact on the earth's natural systems. Paul Ehrlich and John Holdren have proposed a formula that may be viewed as a guide to ecotherapeutic

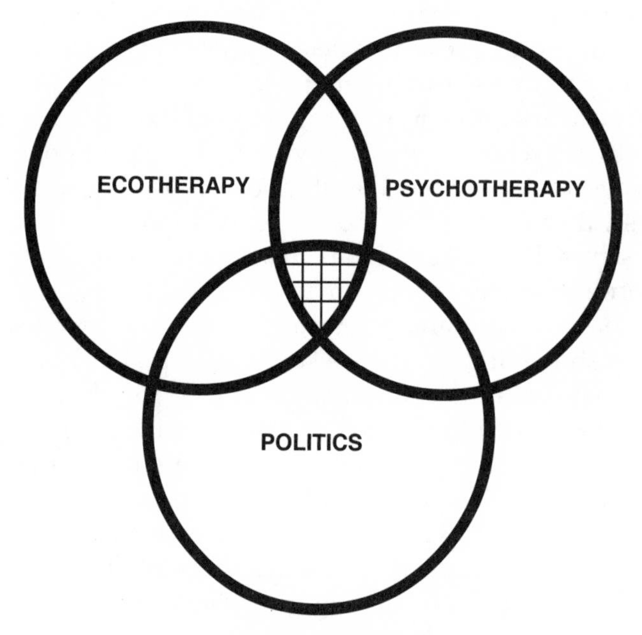
BIOSPHERIC INTEGRITY
(imperative: maximize biodiversity)
PERSONAL WHOLENESS
(imperative: maximize motive satisfaction)
ECOTHERAPY
PSYCHOTHERAPY
POLITICS
SOCIAL COHERENCE
(imperative: maximize social conformity)

Figure 3

practice. The formula is this: $I = P \times A \times T$, where I is a measure of the impact of a human group on natural systems. The variables yielding I are: population size (P), per-capita affluence (A) as measured by consumption, and technological damage (T) done to environmental sources and sinks in the production of items consumed.

This formula tells us that if we wish to reduce human impact on the prospects for viability, then we must act in ways to reduce population levels, consumption levels, and/or technological damage. It makes sense to apply this formula to groups ranging in size from the family all the way up to the level of national economies. The formula allows that different measures may be appropriate for different groups. Population increases in high-affluence groups, for example, would be especially burdensome to natural systems since industrial countries may consume up to three hundred times the per-capita rate of nonindustrials. Sound ecotherapeutic practice in industrial nations indicates a reduction of consumption together with measures to convert to less damaging technologies. Industrial nations should prepare themselves for systematic economic shrinkage. Meanwhile, third world nations should take measures to reduce their populations, but they should also resist temptations to adopt noxious industrial technologies as they seek to modernize.

This stuff is old hat. There is no deep mystery in figuring out the broad ecotherapeutic objectives of a reasonable solution to the global problematique. The measures suggested here have been urged on us for decades by various ecotherapists. Meanwhile, population, consumption, and pollution continue to increase steadily and alarmingly. Our difficulty is less in knowing what to do than it is in finding effective means for getting it done. We are far from knowing all we should about the sound practice of ecotherapy, but we have for some time known enough to set the agenda. The main obstacles are psychological and political.

As we have already observed, there are many ways to become a whole person. But it is not true that all lifestyles conducive to personal wholeness are consistent with a sound ecotherapy. Quite the contrary, it is in the name of personal wholeness that we are driven to behave in ways that have *caused* the global problematique. The culprit in this dynamic has been the wild card of self-esteem. The self-esteem motivator, recall, is aroused by the influence of socially induced values. When certain outcomes are socially valued then individuals who internalize the values become vulnerable to emotional deficit states that can be repaired only by measuring up to the standards. In societies (like ours) where self-esteem is linked to consumptive behavior there will be powerful emotional incentives for excessive consumption. And in societies where self-esteem is linked to family size there will be powerful emotional incentives for excessive reproduction. The linkage of self-esteem to behavioral outcomes is created by social signaling, both positive and negative—that is, by the turned head, the approving glance, the cold shoulder, the snicker, the adjectives chosen, the statusmongery of gossip, and a thousand other subtleties of adulation and reproach. These are the goads and the garlands of self-worth.

The trouble is, of course, that cultures often elicit behaviors that are inconsistent with sound ecotherapeutic practice. We chuckle at frayed cuffs and worn shoes, but we truckle at three-car garages and lavish vacations. But this need not be. The wild card of self-esteem might just as well be linked to behaviors consistent with biospheric integrity as to those conducive to excessive consumption—with no loss of opportunities for self-esteem. Sound psychotherapy under the conditions of the global problematique clearly indicates a need for psychological relinkage. We should scorn opulence and waste while rewarding those who reduce, reuse, and recycle; we should signal admiration for those who get by with less while we spurn those who seek after more. Appropriate psychological relinkage should also be encouraged in parts of the world where excessive population is the critical problem. In these venues large families should be stigmatized, while birth control should be lauded.

There are murmurs of psychological relinkage in a few bright corners of the residual countercultural community, as well as in the promising ecospirituality movement. But we cannot expect the process to develop either quickly or thoroughly without substantial systematic changes in the domain of politics and economics. The reality is that the depth of change required to address the global problematique calls for top-down power as well as bottom-up sentiment. Relinkage is as much a political task as it is a psychotherapeutic one. Consider the case of Christianity. Whatever else the Christian proclamation had going for it, the fact remains that Christianity's biggest breakthrough came with the conversion of the Roman emperor Constantine. One courageous politician can a groundswell make.

Political courage is desperately needed at this hour, for there is much to do. It is no exaggeration to say that conventional political process is deeply flawed, both nationally and internationally. The major problem with traditional political systems is that they have been constructed to address immediate problems with short-term solutions. Long-term issues pertinent to the environment are routinely put aside in favor of more pressing concerns, like stimulating the economy or getting reelected. What makes this a serious flaw is that rational decisions in a short time frame may be palpably irrational in a longer time frame. In other words, what is wise from the short-term perspective of conventional politics is often just plain foolish from the long-term perspective of ecology.

Political process at the international level is equally flawed. Here we find ourselves with an interdependent global economy that has developed far in advance of the institutions required to monitor and regulate it. In other words, we have a global economy without the benefit of a coherent global economic policy, and without decent prospects for getting one. When people share resources and problems (as we do now on a global scale) it is imperative that they have appropriate and effective means for cooperation. Sound politics for our time calls for the creation of a genuine global parliament that can begin to address the challenges of the global problematique. But such a development is precluded by what appears to be an absolute commitment of the world's political leadership to the doctrine of national sovereignty. The flaw here is to suppose that this doctrine can ever be consistent with global solidarity and cooperation.

Sound politics for our time also calls for the creation of bioregional organizations that can monitor local problems and invent local solutions. Solidarity and cooperation at the local level is not rendered insignificant by the need to think and act as a single species at the global level. Just the opposite. Global institutions cannot address all the problems. At best, they can create a political and economic environment in which local communities can experiment with measures for integrating ecotherapy, psychotherapy, and politics.

As matters now stand, our psychological linkages and our social-political-economic organizations are such that they encourage us to act in ways that are inconsistent with biospheric integrity. So what are the options? It is not an option to suggest a redesign of natural systems. This is prevented by the Supreme Court of Biological Realism. We happen to live in *this* biosphere, not in one where the limits can be negotiated to suit our personal wishes and political habits. The only realistic option, therefore, is to adjust the variables of psychological and social organization to conform to the sovereignty of nature. Therein lies wisdom, therein lies hope.

In 1968 the crew of the Apollo 8 flight treated us to the first photograph of earth taken from outer space. That such a photograph was technically possible has a lot to say about the awesome capabilities of our species for understanding and invention. But the deeper significance of the image resides in its power to draw men and women into reflective wonder about its meaning. There hangs a lonely luminous gem in the vast silence of space. There swirl lacy clouds over land and sea, pattern upon pattern. There lie broad sweeps of surf and plain, and aimless massive plates shouldering slowly about. There is a place of boiling and freezing, of torrent and calm. There great orchestral forests tune up with squeaks and howls and chatter. There earthen caves record the wonderings of women and men. There is a place of wounding and healing, of strife and repose, of living and dying, and again living.

How did this busy blue sphere come to be? What is it doing there? Has it any enduring promise? What is the story of this storied place? It is perhaps naive to pretend that we could ever finally know. But it is feckless not to ask, not to seek, not even to hope that some understanding might be earned. The photo from space has taught us one thing for sure: *there is only one story*. We are forever beyond the multiple stories of this mountain or that valley, of this or that tribe or nation or god. From here onward, our longing is for everybody's story. We know this too: everybody's story is not a single find, not a once-and-for-all-time revelation that downloads complete to solitary prophets. It is a shared, disciplined, debated thing—never final, never orthodox, never completely true. But it is our best chance to behold an enduring promise. Everybody's story needs many voices and many versions, but if it is to be everybody's story then those venturing to tell it must stand out there, at some distant remove, where the earth can be seen whole.

Epilogue

The Prospects for Wising Up

This little book has lofty aspirations: to participate in the most important intellectual endeavor of the new millennium—that is, to stimulate the emergence of a new wisdom tradition based on the integration of evolutionary cosmology and ecocentric morality. This new story, everybody's story, is full of potential for uniting our species around a common apprehension of how things are and which things matter. United by a shared story, we may come to possess a sense of solidarity and cooperation sufficient to inspire the will to seek human fulfillment within the limits of biospheric integrity.

So goes the rosy scenario. But really, what are the prospects? Dreams of uniting humanity around a common set of ideas are a dime a dozen, and as old as the hills. What makes the odds for this one better than zero? What is there to justify confidence in the potential of everybody's story?

For one thing, the severe nature of the challenge will keep the pressure on to wise up. There is nothing superficial or ephemeral about the global problematique. It is global, systematic, immediate, and chronic. And it will surely worsen so long as human population and consumption continue to rise on a global scale. If there is anything remotely positive about the global problematique it is only that it has brought us to a serious encounter with our limitations as a species. Like the experience of pain, sensing our limits is unpleasant but instructive. Each encounter with human limitations is an invitation to wise up to a deeper self-understanding. The word has gone out that viability is at stake, and people are beginning to assess the options.

My confidence in the potential of everybody's story to stimulate a new wisdom tradition is based, in part, on a belief that we will continue to make progress in science. I may be mistaken in this belief, for it is possible that we will witness a sudden reversal in the status and influence of science. There are serious antiscience voices heard in public places these days, and there may be some reason for concern that suspicion of science will broaden and intensify. But I doubt it. More worrisome is the possibility that a greater share of the scientific community will be coopted by the zealots of unrestrained economic growth. If the captains of industry are left to direct progress in science, then research and development will become even more consumer driven than ever.

But even in this sorry event, there will be enough scientific expertise on hand to give us ever more evidence of the seriousness of our environmental problems. If the industrialists want good scientists then there will have to be an adequate educational infrastructure to train them. But once trained, good scientists, as a group, are notoriously difficult to control politically. They have this thing about the truth. And if the truth seems to be that overpopulation and overconsumption are undermining the integrity of natural systems and threatening viability, then the scientific community can be counted on to say so.

If progress in science is allowed, then there will continue to be pressure for increasing scientific literacy. These two things—progress in science and increased scientific literacy—give us a formula for advancing the potential of everybody's story. To the extent that science progresses, the features of everybody's story will become more complete and more compelling. And to the extent that scientific literacy increases, there will be ever more individuals compelled by it.

As long as science education is on the screen, I might as well toss in a prediction about the way science will be taught in the future. It will not be long before science educators realize that the brain is a narrative spinning modular system. As soon as they do, we can expect to see the sciences taught in a narrative format. Kids will learn more science, and remember it better, when it is delivered to them as a story about emergent systems. My prediction is that the science curriculum of the future will have the broad contours of everybody's story.

A wildly optimistic scenario might take the idea even deeper, to the level of the collegiate core curriculum. Whatever challenges are faced by the college graduates of the next century, they must bring to bear on these problems an adequate understanding of human nature. Having an adequate understanding of human nature means having an integrated comprehension of natural and social systems. In recent years university faculties have been stalled in a misguided debate over the substance of the core curriculum. On one side are the voices of tradition, defending a privileged status for certain classics of the Western cultural experience. On the other side are the voices of diversity, insisting that there are many more ways than one to be authentically human, and that the core curriculum should be multicultural. It has really been a debate over which stories should occupy the attention of students during the formative college years. Far superior to either of these options would be a core curriculum focused on the evolution of matter, life, and consciousness-culture. If we begin to see such a development, then we may be confident that the root system of a new wisdom tradition has taken hold.

Every wisdom tradition is filled with confidence that its story has the power to arouse the imagination and compel the will to follow its imperatives. The important thing is to get the word out there, where it can do its work. But how does a story do its work? What are the mechanics by which it compels the will?

It is, in the end, a matter of motivation. It is not enough to know how things are, nor is it enough merely to know the good. The Greek philosophical tradition simply had it wrong in assuming that knowledge is sufficient to wisdom. To act on what we know requires that we also feel. But passion is not sufficient either. Ignorance in motion is dangerous, unmoved knowledge is impotent. True wisdom can be achieved only by well-informed motives. My confidence in everybody's story to unite humanity rests on my belief that this story can do more than satisfy our longings to know, it also has the potential to arouse and direct the emotional regulators of behavior.

Consider the following examples taken from ancient wisdom traditions. The Hebraic tradition emphasizes the majesty and awesome power of God, the creator of the universe and ultimate judge of human righteousness. Almighty God enters into a covenant with Israel, promising viability and prosperity on the condition that Israel obey his commands. Blessings if they do, suffering and hardship if they do not. God's people are made to feel humble and awestruck in the presence of absolute power and authority. They are made to feel grateful for the undeserved bounty from God, and duty-bound to observe his Law. The logic of reciprocity could not be more obvious. The imagery of the story activates the social emotions, creating a deficit state that seeks repair in obedient service. Similarly, see how the central images of the Christian story are designed to arouse the social emotions. *Affection* is elicited by images of the infant Jesus, mother and child, the caring shepherd; *sympathy* is aroused by the image of a helpless and innocent man suffering on a cross at the hands of merciless authorities; *gratitude* is called forth by the reminder that Jesus' death was a selfless act undertaken for the sake of others; *guilt* is instilled by the insinuation that it is we who deserve the punishments of the cross; *resentment* or *moral outrage* is aroused against those, like Judas, who betray the altruistic Jesus. By such imagery the emotional effectors are set to work in motivating a life of service to Christian virtues.

There is nothing in the substance of everybody's story to rule out belief in the reality of a personal deity. At the same time, such a belief is not an essential *part* of everybody's story. There will be theistic versions of the story, and there will be nontheistic versions as well. Those who take the theistic option will have at their disposal a range of images that may be used to arouse motivational systems. But I have confidence that everybody's story, unadorned by theological imagery, has the potential to arouse us to serve its imperatives. Let us see.

The social emotions have been endowed upon us by natural selection because of their role in mediating biologically advantageous cooperative behaviors. Nevertheless, these emotions are quite commonly and naturally aroused in us apart from social contexts. Imagine yourself alone in the woods, gathering fuel for the fireplace. As you pass through the trees a massive bough breaks loose above and comes crashing down, missing your head by mere inches. Pure gratitude. Or perhaps you encounter a helpless fawn with its leg wedged between a rock and a hard place, unable to escape. Unmitigated sympathy. Or you come across a mare, licking the cheek of her newborn foal. Flat-out reverence. Or you trip over a concealed log as you make your way through the brush. Your immediate response is to resent the log, perhaps intensely enough to curse at it.

These are perfectly natural responses, just what is expected from an organism equipped with our emotional repertoire. Likewise, when one is brought to reflect on the events of everybody's story—the immensity, grandeur, and improbability of the universe; the fortuitous position of our planet; the delicacy of the ozone shield; the billions of random mutations; the miracle of multicellularity; the splendors of biodiversity; the excitement of the first neuron—when one is brought in reflection to these events the social emotions go berserk. It just happens, no imagery required. The unvarnished narrative account is sufficient. That I could have the life I do, at the end of all this interplay of chance and necessity, is just too much to fathom. If I listen, and if I then reflect, I cannot remain still. Gratitude abounds.

Gratitude is, of course, a deficit state, it needs repair. And here the gods were born. Imagine our earliest ancestors who were spared in battle, who made it through a harsh winter, or a harrowing flood. Overwhelmed with reflective gratitude, they found a way to give thanks. How do you behave when you receive an unsigned gift? It is frustrating. So you begin to imagine who might have sent it. The events of everybody's story leave us feeling gifted, burdened with gratitude, in need of expression.

Everybody's story needs imagery too. Not to arouse a sense of gratitude for the lives we have—the unvarnished truth, goodness, and beauty of the epic of evolution are more than sufficient for that—but to satisfy it, to express it. What should the images be? What imagined target for our gratitude would best direct the will to serve the imperatives of the story? Again, theists may draw on the imagery of divine grace, directing their sense of gratitude through traditional conventions. But nontheists are at no disadvantage here. For nontheists it will be sufficient to imagine a future. If thanksgiving requires a face, then let it be the face of future generations. Never mind that thanking the future appears to violate the logic of reciprocity (what has posterity ever done for us?). That should not be a problem. In fact, there are worthy precedents for such redirection of gratitude in the ancient wisdom traditions themselves. Jesus bade his followers, if they loved *him*, to feed his *sheep*. The prophet Amos portrays God's disgust at elaborate festivals of thanksgiving, preferring instead that justice prevail in the courts. This sort of redirection of social emotion is appropriate to our story. If the ultimate value is the continuation of life, then it makes perfect sense to repay our gratitude for the evolutionary past by endowing the future. Everybody's story spawns not an ancestor cult, but a descendant cult.

The imagery allowed to develop within a cultural tradition is no insignificant matter, for it is imagery that educates the emotions and directs the will. Look at the images dominating a cultural tradition and you will find insight into its character, its organization of consciousness. The popular culture of medieval Europe was dominated by religious symbolism. Everywhere one looked, the Christian story was reinforced by imagery. But what do we find today? If visitors from space were to analyze contemporary industrial culture they would surely conclude—and rightly so—that we are a cult of consumption. We are saturated by images that arouse greed and direct our wills to the nearest shopping mall. There is virtually nothing in the symbolism of popular culture that allows us (not to say compels us) to imagine our connections with the past or with the future. We are stranded on an island of immediacy. Everybody's story invites us to surround ourselves with the symbolism of natural history, an array of images that will arouse gratitude for the epic of evolution and transform this deficit into a commitment to future generations. The images have not yet arrived, but I am confident they will come—if the story gets out there, where it can do its work.

And what if it does? What then? Will we then have a new religion on our hands? Is that ultimately what is being proposed in this book—that everybody's story is a candidate for everybody's religion? I hesitate.

The word "religion" literally means that which "binds together." And it certainly has been my claim that everybody's story provides a means for binding humanity together into a global culture. Everybody's story gives us a shared perception of how things are and which things matter. It orients us in nature and in history. It tells us in fundamental terms who we are, whence we have come, and how we may be fulfilled. It has the potential to inspire gratitude and direct the will to serve ultimate values. If that is not religion, then what is it? Still, I hesitate.

I have no doubt that for a substantial number of people the features of everybody's story will enlarge to fill all the roles of their former religious orientations, without remainder. Thus in time we will see the development of new ancillary strategies designed to transmit and maintain particular versions of everybody's story. Earth Day, for example, may eventually acquire all the ritualized pomp and ceremony of Easter. It is almost a certainty that various forms of nature mysticism will arise—indeed, they are already with us. And of course a new aesthetic tradition will emerge as those with talent for creative expression seek to repair the emotional deficits aroused by everybody's story. But there will always be a substantial number for whom everybody's story leaves a remainder—those for whom the epic of evolution does not say all they want to hear. Even after everybody's story has been heard, there will be many whose needs for intelligibility, hope, and emotional fulfillment are left unsatisfied. For these individuals, everybody's story will not have the feel of a religious orientation. This is not to say that the epic of evolution will be excluded from their religious lives, it merely means that everybody's story is not for them the whole story.

So I hesitate to characterize everybody's story as a religion. For some, it may become so, but for others it will not. In any event, it does not pretend to be everybody's religion. I prefer to use the term "wisdom tradition" as a label for this story. It will be enough for everybody's story to become a bedrock of global wisdom, to be a common theme tolerating a diversity of interpretations—in the same way that our common humanity tolerates our peculiar biographies. If things get this far, then we shall have the means for global solidarity and cooperation, and everybody's story will have done its work.

Finally, I take confidence in the prospects for everybody's story from my judgments about the future of religion. The future of religion has been a hot topic ever since the Enlightenment of the eighteenth century. For much of this period the critical question has been whether religion can even expect to *have* a future. Freudian, Marxist, and positivist claims about the eventual demise of all religion are well known. More recently, however, opinions about the future of religion have changed. No one these days is predicting the end of religion. Religion, like politics, is now generally accepted as a universal human phenomenon. Wherever there are human communities, there you will find religion in one form or another. Nevertheless, there are many who expect particular religious and theological traditions to fade away. For example, one might seriously question whether the Axial traditions have much to look forward to. The real question, then, is not whether religion will persist in the future, but rather what it will be like.

One thing, at least, is clear. As the global problematique worsens, and as scientific literacy expands, the received religious traditions of the world will continue to experience a deepening crisis of intellectual plausibility and moral relevance. Some interpretations of the Axial traditions are simply impossible to reconcile with contemporary cosmology. Likewise, some interpretations of these traditions are incommensurate with the moral challenges imposed by the global problematique. These circumstances generate what looks like a no-win situation for the Axial traditions. On the one hand, if the Axial traditions fail to assimilate the evolutionary cosmology and the ecocentric morality of everybody's story, then the dual crisis of plausibility and relevance will deepen within these traditions to accelerate their decline. On the other hand, if the Axial traditions succeed in absorbing everybody's story they risk compromising their distinctiveness to the point of appearing superfluous. This dilemma is already acute in the case of Christian theology. The scientifically literate have long since written off fundamentalism as unbelievable and immoral. At the same time they are not convinced that there is a difference between ultraliberal theology and no theology at all.

So what are the options for the received traditions? The fundamentalist option is unrealistic and dangerous. It tends toward antiscientism and cultural chauvinism. Fundamentalism is no strategy for global solidarity and cooperation. The only alternative open to the Axial traditions is to reinterpret their symbols and doctrines in ways that are consistent with evolutionary cosmology and ecocentric morality. Many Axial theologians are already busily involved in this very important work. They are convinced that a theological reinterpretation can proceed along these lines without a loss of distinctiveness. More power to them, for in the process of assimilating everybody's story they will succeed in advancing its potential. This potential is to construct a common and minimal framework of wisdom within which a broad range of spiritual practices might prosper. Philip Hefner employs the image of loom and fabric. The epic of evolution provides the basic frame of reality and value upon which many tapestries of meaning may be woven. That Hefner's image may one day be a fair description of religious practice throughout the world is the underlying hope of this book.

Despite the no-win plight of the Axial traditions, one may expect a vigorous future for alternative forms of the religious life. As the potential of everybody's story continues to unfold, more and more women and men will come to define themselves religiously as *seekers*, that is, as natural beings engaged in a continuous search for a more satisfying relationship with Reality. In essence, of course, the religious life has always been about seeking. Religious traditions, at their best, have consistently seen themselves as frameworks for spiritual striving and growth. The great saints were always on the move, always wising up.

The most fundamental lesson of the evolutionary process is that evolution never starts from scratch, it works with what it finds. We always begin precisely where we are. We are, at the moment, in many different places, with many histories and hopes. But we are now called together to one place, to a shared history and to a common vision of enduring promise. If there are saints enough among us we shall arrive.

Bibliographical Notes

General

Chaisson, Eric (1981). *Cosmic Dawn.* W. W. Norton.
Follows the arrow of time from the big bang to the emergence of technological culture.

——— (1987). *The Life Era.* W. W. Norton.
Draws the big picture of cosmic evolution, with special focus on the emergence of matter from radiation, and life from matter.

Cloud, Preston (1978). *Cosmos, Earth, and Man.* Yale Univ. Press.
The epic of cosmic and biotic evolution by a consummate geologist.

Layzer, David (1990). *Cosmogenesis.* Oxford Univ. Press.
A sometimes technical account of emergent order in the expanding universe—from creation to consciousness.

Scientific American. Special issue on "Life in the Universe." (October 1994).
Informative articles by Carl Sagan, Stephen J. Gould, Steven Weinberg, Marvin Minsky, and others, telling the story from the big bang to the global problematique.

Swimme, Brian, and Thomas Berry (1992). *The Universe Story.* HarperCollins.
A mathematical cosmologist and a historian of culture team up to tell the story of cosmic evolution.

Introduction: The Challenge of Wising Up

Alexander, Richard (1987). *The Biology of Moral Systems.* Aldine de Gruyter.
A sociobiological account of moral behavior, including theories of kin selection and reciprocal altruism.

Barney, Gerald (1993). *Global 2000 Revisited*. The Millennium Institute.
A concise and competent survey of the global problematique.

Brown, Lester, *ed.* (1995). *State of the World*. W. W. Norton.
An annual report from the Worldwatch Institute on various aspects of the global problematique.

Ehrlich, Paul, and Anne Ehrlich (1991). *Healing the Planet*. Addison Wesley.
A readable account of nature's life-support systems, and how they are being stressed by human population and consumption.

Eisenstadt, S. N., ed. (1986). *The Origins and Diversity of Axial Age Civilizations*. SUNY Press.
A variety of perspectives on the character of axial traditions, and the conditions out of which they emerged.

Ponting, Clive (1991). *A Green History of the World*. Penguin Books.
A grim history of humanity's interactions with nature.

Service, Elman R. (1978). *Profiles in Ethnology*. 3rd ed. Harper & Row.
An anthropological survey of patterns of human social organization.

Chapter 1—The Organization of Matter

Cesare, Emiliani (1992). *Planet Earth*. Cambridge Univ. Press.
For serious readers looking for the details of contemporary cosmology, geology, and the evolution of life.

Cloud, Preston (1988). *Oasis in Space*. W. W. Norton.
A complete and authoritative history of the earth, from its origins to the arrival of humans.

Davies, Paul, and John Gribbin (1992). *The Matter Myth*. Simon & Schuster.
A popular account of how contemporary physics is changing our understanding of reality. Davies and Gribbin have each authored several other books relevant to the organization of matter.

Hawking, Stephen (1988). *A Brief History of Time*. Bantam Books.
Reflections on the early universe by one of this century's leading cosmologists.

Jastrow, Robert (1969). *Red Giants and White Dwarfs*. Harper & Row.
A classic survey of cosmology, geology, and the origins of life. A bit dated, but still worth reading.

Silk, Joseph (1980). *The Big Bang: The Creation and Evolution of the Universe*. W. H. Freeman.
A leading cosmologist summarizes prevailing theories of the universe for a general audience.

Weinberg, Steven (1988). *The First Three Minutes*. 2nd ed. Basic Books.
A Nobel laureate physicist gives an accessible account of contemporary cosmology.

Chapter 2—The Organization of Life

Crick, Francis (1981). *Life Itself: Its Origins and Nature*. Simon & Schuster.
Just what the title promises—from the codiscoverer of DNA.
Dawkins, Richard (1987). *The Blind Watchmaker*. W. W. Norton.
A beautifully written exposition and defense of Darwin's theory of evolution by blind natural selection.
Dyson, Freeman (1985). *Origins of Life*. Cambridge Univ. Press.
A first-rate mathematician and theoretical physicist (and a fine writer) examines several theories of life's origins.
Lovelock, James (1988). *The Ages of Gaia*. W. W. Norton.
A presentation and defense of the Gaia hypothesis, the view that life contributes to the self-organizing geochemical processes that favor its continuation.
Margulis, Lynn, and Dorian Sagan (1995). *What Is Life?* Simon & Schuster.
Mother-son duo team up to produce a visually and intellectually satisfying response to the questions of life's origins and nature.
Wilson, Edward O. (1992). *The Diversity of Life*. W. W. Norton.
Already the standard text on biodiversity, by one of the world's most eminent biologists.

Chapter 3—The Organization of Consciousness

Calvin, William (1990). *The Ascent of Mind*. Bantam Books.
A neurobiologist narrates the evolution of intelligence against a backdrop of climatic and environmental change.
Damasio, Antonio (1994). *Descartes' Error*. Grosset/Putnam.
A noted neurologist makes a coherent and well-documented case for the fundamental role of emotion in human rationality.
Dennett, Daniel (1991). *Consciousness Explained*. Little Brown.
An engaging account of consciousness as a consequence of evolved subconscious physical processes.
Diamond, Jared (1992). *The Third Chimpanzee*. HarperCollins.
A provocative account of the rapid evolution of human behavior, and the potential dangers that lie ahead.
Ornstein, Robert, and Paul Ehrlich (1989). *New World New Mind*. Simon & Schuster.
An account of how human intelligence has been organized by evolution, and what this means for addressing contemporary global problems.
Sagan, Carl (1977). *The Dragons of Eden*. Ballantine Books.
A well-written introduction to the evolution of intelligence—a bit dated by the fast pace of brain research, but still worth reading.

Chapter 4—What Matters Ultimately?

A few good books pertinent to evolution and morality:

Alexander, Richard (1987). *The Biology of Moral Systems*. Aldine de Gruyter.
Bradie, Michael (1994). *The Secret Chain: Evolution and Ethics*. SUNY Press.
Cronin, Helena (1991). *The Ant and the Peacock*. Cambridge Univ. Press.
Darwin, Charles (1982). *The Descent of Man and Selection in Relation to Sex*. Princeton Univ. Press.
Dawkins, Richard (1976). *The Selfish Gene*. Oxford Univ. Press.
Midgley, Mary (1979). *Beast and Man: The Roots of Human Nature*. Harvester Press.
Moore, G. E. (1962). *Principia Ethica*. Cambridge Univ. Press.
Rachels, James (1990). *Created From Animals: The Moral Implications of Darwinism*. Oxford Univ. Press.
Ruse, Michael (1986). *Taking Darwin Seriously*. Blackwell.
Wilson, Edward O. (1984). *Biophilia*. Harvard Univ. Press.

Chapter 5—What Matters Proximately?

A few good books pertinent to ecocentric morality.

Bookchin, Murray (1990). *Remaking Society: Pathways to a Green Future*. South End Press.
Daly, Herman (1977). *Steady-State Economics*. W.H. Freeman.
Leopold, Aldo (1981). *A Sand County Almanac*. Oxford Univ. Press.
Naess, Arne (1989). *Ecology, Community, and Lifestyle*. Cambridge Univ. Press.
Roszak, Theodore (1992). *The Voice of the Earth*. Simon & Schuster.
Sale, Kirkpatrick (1985). *Dwellers in the Land*. Sierra Club Books.
VanDeVeer, Donald, and Christine Pierce, eds. (1994). *The Environmental Ethics and Policy Book*. Wadsworth.

Index

algae, 100
Amos, 134
Apollo 8, 126
Aquinas, Saint Thomas, 36
Aristotle, 36, 112
art, 25–26
atmosphere, 60
Augustine, Saint, 36
Axial age, 28–34; chaos of, 30
Axial traditions, 29–39; cosmology of, 31–32, 34; crisis in, 34–39, 136–37; influence of, 34–39, 136; morality in, 30–34

bacteria, 71
Barney, Gerald, 3
big bang, 51–52, 56, 62, 81
biodiversity, 101, 121–22, 133
bioregionalism, 126
biosphere, 60; integrity of, 121–26
black holes, 58
brain, functions of, 22, 75, 82–84, 91; modularity of, 90–91, 94, 131
Buddhism, 29, 38

central nervous system, 22, 49, 82, 84, 91
chiefdom, 15–17, 32
China, 29
Christianity, 29, 35–36, 38, 101–102, 132, 134
conflict, psychological, 116–18
Confucius, 29
colonialism, 8–9, 16
consciousness, conditions for, 82–84; emergence of, 47–49, 81–84; global, 29; nature of, 81–82; organization of, 84–96; transformation of, 2, 18–20; *See also* transformation
consensus; ancillary strategies for, 25–28
consumption, excessive, 6–7, 15, 17–20, 121, 124, 130
cooperation. *See* solidarity and cooperation
cosmic evolution, 43, 47–52. *See also* evolution; universe
cosmology, 23–27, 136; Axial, 31–39; evolutionary, xiii, 129, 137; *See also* how things are

crisis, of plausibility. *See* intellectual plausibility; of relevance. *See* moral relevance
culture, defined, 22–23, 87; diversity of, 89, 119; global, 29; learning and, 87–88; proto vs. symbolic, 88; structure of, 22–23, 87

Damasio, Antonio, 93–94
Darwin, Charles, 105–106
Dawkins, Richard, 76, 88
death, 32, 66, 79–80
determinism, 55
divine command theory, 103
divine purpose, 63–64, 75–78
DNA (deoxyribonucleic acid), 68–69, 71–72, 74, 77–79
domestication of animals, 15
dualism, 31–33, 37, 48

Earth Day, 135
Easter, 135
economic growth, 6–7, 130
ecospirituality, 125
ecotherapy, 121–26
Ehrlich, Paul, 121–22
emotions, 10–14, 18, 113–16, 124, 132–34
energy transformation, 67, 77
Enlightenment, 35
entropy, 76–77
epic of evolution, xii–xiv, 135, 137. *See also* everybody's story; evolution
eukaryotic cells, 71
everybody's story, 28, 41, 48–50, 89, 120, 127, 129–37; and imagery, 134–35; potential of, 129–30, 137; and religion, 132, 135–36; as wisdom, 129–32, 136; *See also* epic of evolution
evolution, cultural, 88–89; cosmic, 43, 47–52; of intelligence, 78; paradigm of, 42–43; of universe, 42–43; *See also* epic of evolution
existentialism, 90

folly, ecological, 3–7, 20
foundationalism, moral, 105
Freudianism, 136
fundamentalism, 137
future, xiii, 49, 134–35

galaxies, 50–51, 57–59
genetic code, 69, 106
genetic variation, 69
germ line, 72, 79–80
global problematique, 3–10, 17–18, 28–29, 34, 37, 129–30, 136
Gore, Albert, Jr., xi
gratitude, xii–xiii, 49, 61, 80, 92, 132–35
gravity, 50–51, 57–59, 62
Greece, 29

Hefner, Philip, 137
heredity, 69
Hinduism, 29, 38
Holdren, John, 121–22
how things are, ideas about, 22–24, 27–28, 30, 45–96 passim; *See also* cosmology
human nature, 37–38, 89–91, 120
Hume, David, 103–104
hunker down option, 9–10, 13, 15
hydrosphere, 60

$I = P \times A \times T$, 123
idealism, 48
improbability, of existence, 49, 62–64, 133
indeterminacy, 55
India, 29
individualism, 32–33, 37
information processing, 67, 77–78, 84–85, 90–91

intellectual plausibility, 28; of Axial traditions, 34–39, 136–37
Islam, 29, 38
Israel, 29

Jainism, 29
Jesus, 134; as divine being, 35
Judaism, 29, 38, 132, 134

kill-off option, 9, 15
kin selection, 11–13, 18, 49
kinship bands, 12–14

language, 87–88, 90, 94
Lao-Tse, 29
learning, 86–88
life, definitions of, 65–67, 100; diversity of, 71–73, 75; emergence of, 47–48, 65, 67; functions of, 65–67, 77–78; meaning of, 101–102; organization of, 67–80; and valuation, 99–101
lithosphere, 60

Marxism, 136
materialism, 48–49
matter, nature of, 53–56; organization of, 56–64; properties of, 48, 54–57, 60–61, 84
meiosis, 72
meme, 88–89, 102
metabolism, 67–71
mitosis, 72
Moore, G. E., 103–104
moral consciousness, 104–105
morality, 23–27, 136; Axial, 30–39; ecocentric, xiii, 129, 137; *See also* values; which things matter
moral reasoning, 104–105
moral relevance, 28; of Axial traditions, 34–39, 136–37
motivational systems, 110–20; curiosity motivators, 110–13, 115–16; hedonic motivators, 110, 112–13, 115–16; social motivators, 110, 113–16, 132–34
multicellularity, 72, 79, 133
multiculturalism, 40–41, 131
mutations, 69–70, 74–75, 133
mysticism, 135
myth, 23, 25–26. *See also* story

narrative; core of culture, 22–28; *See also* story
naturalistic fallacy, 103–105
natural selection, 69, 75, 85–86, 90
natural systems, threats to, xiii, 3–5, 7–8, 20–21, 30, 37, 130; and values, 120–21
nature-nurture debate, 89–91
neo-Darwinism, 90
neo-Platonism, 36
neural self. *See* self, neural
neural systems, 82–87, 90–94
new world order, 28–29
nihilism, 105

Orphic tradition, 29
ozone shield, 70–71, 73, 133

Pangaea, 60
parasitism, 68
Persia, 29
personal success, 6–7, 25
personal wholeness, xiv, 117–26. *See also* social coherence
philosophy, 25–26, 29, 33, 131
photosynthesis, 70, 78
pleasure principle, 112, 115–16
pluralism, 40–41
politics, 121–26
Poorlandia, 4–5
population, excessive, 6–8, 15, 17–20, 121, 124, 130
positivism, 136
prokaryotic cells, 71
protein synthesis, 68–69, 74
psychotherapy, 121–26

quantum theory, 42, 55

religion, xiv, 25–26, 33, 102, 135–37
reciprocal altruism, 11–13, 18, 49
RNA (ribonucleic acid), 68–69, 74
replication, 67
reproductive success, 101
respiration, 70
reverse engineering, 76–77

Sacks, Oliver, 1
salvation, 32–33, 96
self, Buddhist view of, 93–95; Cartesian view of, 93–95; and narrative, xi, 93–94; neural, 93–95
self-esteem, and consumption, 7, 18–19; as motivator, 114, 119, 124–25
sexual reproduction, 72–74
sit-back option, 8, 15
science, integration of, 41–43; literacy, 130–31, 136; as narrative, 41–43; rise of modern, 35–36; philosophy of, 36; progress in, 130
social coherence, xiv, 117–26. *See also* personal wholeness
social organization, means of, 10–13, 118; types of, 10–17
social progress, 6–7, 25
social systems, threats to, 3–5, 7–8, 20–21, 30, 37
Socrates, 3
solar system, 59
solidarity and cooperation, 2; in Axial traditions, 33–34; resources for, xiii–xiv, 10–18, 21–28, 30, 49, 114, 121
soma line, 72, 79–80
spread-out option, 8–9, 12, 16, 89
stars, 50–51, 58–59
state system, 16–17, 19, 30
story, importance of, 20–28, 88–89; new, xiii–xiv, 30, 34, 39–43, 129; profound sense of, 22–23; as source of gratitude, xii–xiii; universal vs. particular, 40–41; *See also* narrative
storytelling, xi, 14, 21, 34
supernovae, 58–59
Supreme Court of Biological Realism, 126
symbols, 13–16, 18, 22–25, 87, 134–35

transformation, social and psychological, 18–21, 29, 34
tribal alliance, 13–16, 30, 32

universe, 50–52; age of, 51; destiny of, 51–52; evolution of, 42–43; expansion of, 51–52, 56–57, 62, 64; utility function of, 76–77
utility function, 76–79

values, and life, 100–101; proximate, 120–26; ultimate, 99–107, 109–10, 120; *See also* morality; which things matter
vascular systems, 85
viability, and morality, 105–106; as objective value, 103–105; threatened, 7, 21, 123, 130; as ultimate value, 101–102, 106, 109–10, 120; as universal value, 100–102
vitalism, 48

which things matter, ideas about, 22–24, 27–28, 30, 99–127 passim; proximately, 109–27; ultimately, 99–107; *See also* morality
wise-up option, 10, 12–15, 17–21, 28
wisdom, 15; in Axial age, 29–33; and culture, 23; meaning of, 2, 21, 132; traditions, xiv, 39–40, 130–36

Zoroastrianism, 29

ADIRONDACK COMMUNITY COLLEGE
3 3341 00074290 7